Modeling Reality

Modeling Reality

How Computers Mirror Life

Iwo Białynicki-Birula

Center for Theoretical Physics, Warsaw, Poland

Iwona Białynicka-Birula

Galileo Galilei School, University of Pisa, Italy

OXFORD

UNIVERSITY PRESS

OXFORD
UNIVERSITY PRESS

Great Clarendon Street, Oxford OX2 6DP

Oxford University Press is a department of the University of Oxford.

It furthers the University's objective of excellence in research, scholarship,
and education by publishing worldwide in

Oxford New York

Auckland Bangkok Buenos Aires Cape Town Chennai
Dar es Salaam Delhi Hong Kong Istanbul Karachi Kolkata
Kuala Lumpur Madrid Melbourne Mexico City Mumbai
Nairobi São Paulo Shanghai Taipei Tokyo Toronto

Oxford is a registered trade mark of Oxford University Press
in the UK and in certain other countries

Published in the United States
by Oxford University Press Inc., New York

A catalogue record for this title is available from the British Library

Library of Congress Cataloging in Publication Data

Białynicki-Birula, Iwo.
   Modeling reality : how computers mirror life / Iwo Białynicki-Birula,
Iwona Białynicki-Birula.
      p. cm.
   Includes index.
   ISBN 0-19-853100-1 (alk. paper)
   1. Reality. 2. Life. 3. Physics—Philosophy. I. Białynicka-Birula, Iwona. II. Title.
   QC6.4.R42B53 2004
   530′.01—dc22
                                                                    2004018291
ISBN 0 19 853100 1 (hbk)

10 9 8 7 6 5 4 3 2 1

Typeset by Newgen Imaging Systems (P) Ltd., Chennai, India.
Printed in Great Britain
on acid-free paper by Biddles Ltd., King's Lynn

We dedicate this book to Zofia Białynicka-Birula

Preface

This book originated from a series of lectures delivered by the first author at the Warsaw School of Social Psychology and at Warsaw University over the last six years. The purpose of these lectures was to give a very broad overview of various aspects of modeling for a mixed audience, from students of mathematics, computer science, and physics to students of biology and social sciences. Considering the different levels of mathematical literacy among those who attended the lectures, we have relied only on the mathematical concepts known to high school graduates. Therefore, our book can be understood by a wide spectrum of readers—from ambitious high school students to graduate students of all specialities. We were trying to keep the mathematics at the high school level; however, some chapters may require an additional effort since they describe modern advances in computer modeling.

The book was originally published in Polish in 2002 by the publishing house Prószyński and Ska in Warsaw. The English translation is substantially expanded and modified.

The material presented in this book is illustrated with twenty-five computer programs written especially for this purpose. All of the programs have undergone a substantial upgrade and face-lift for the English edition.

Acknowledgements

We warmly thank Vittorio Bertocci for many valuable comments and suggestions regarding both the text of the book and the form of the programs. This work has been partly supported by the Polish Ministry of Scientific Research Grant Quantum Information and Quantum Engineering.

All portraits of the scientists and the chapter opening images were drawn by Maria Białynicka-Birula.

Contents

From building blocks to computers

Models and modeling

The aim of science is to describe, explain, and predict the behavior of the world which surrounds us. Reality is, however, much too complex to be described accurately, without any simplification or approximation. That is why, when describing a given phenomenon, we take into account only the elements of reality which we think have a significant influence over the phenomenon we are describing. Let us suppose, for example, that we are describing the motion of an apple falling from a tree. To describe it we would take into account the weight of the apple, the height of the tree, and perhaps the air resistance. On the other hand, we would omit such factors as the taste of the apple, its color, or the configuration of the planets in the sky at that given moment. The planet configuration does in fact influence the motion of the apple, but that influence is so small that we can easily omit it.

Specifying the significant factors is perhaps the most important element of research. Many fields of science, especially astronomy, physics, and chemistry, started developing rapidly only when the scientists learned to recognize the significance of different factors and to omit the insignificant ones. In other fields, such as social science, psychology, or medicine, the significance of different factors has not yet been precisely established.

Conducting research in a given field can be made easier if one first tries to understand the behavior of very simple objects, hoping to thus shed light on rules governing the behavior of the more complex ones. The concepts and methods covered in this book can considerably aid such an approach.

The word **model** appeared in written English for the first time in 1575. Nowadays it has many meanings. In Webster's *New encyclopedic dictionary* we find nine and in *Merriam–Webster dictionary* even thirteen different meanings of the term *model*. In a recent book (Müller and Müller, 2003), the authors give nineteen illuminating examples of different uses of this notion. There are two meanings listed in Webster that are perhaps the closest to our usage of this term. These are: 'a system of assumptions, data, and inferences used to describe mathematically an object or state of affairs' and 'a theoretical projection of a possible or imaginary system'. In this book, by a **model** we shall mean the set of elements of reality considered to be significant for a given phenomenon and the rules governing the behavior of these elements. The choice of significant elements and the definition of the rules is indeed the essence of modeling. We can judge the suitability of this choice by comparing the results obtained from the model with reality.

From cradle days, every one of us comes across many different models. Even the cradle itself is a model. Most children's toys are models of objects from the adult world and games are models of real social situations. Our adult life is full of models as well. A map is a model of a city or a country, a globe is a model of the Earth, and a calendar is a model of a year. When conducting a poll, as a model of society, we often use a small sample of its representatives. A computer is a model of evolution's grandest creation—the human brain. All computers can in turn be modeled by a single object, called the Turing machine, an abstract device described in detail later in this book. Even the human intellect can be modeled, its model being what is called artificial intelligence.

The act of modeling can be split into three phases. The first is the most difficult of all as it does not conform to any precise rules—it is **choosing the model** to describe a given aspect of reality. The second phase is **constructing an algorithm** according to which the model will function. An algorithm is a set of rules, which when applied systematically leads to a solution to the problem. The third phase, easiest, though often arduous and time-consuming, is checking the obtained results against the initial hypothesis and **drawing conclusions**. Naturally, in the end it is the conformity of the results with the reality that we meant to model that serves to measure the model's worth.

Algorithms were known to ancient Greek mathematicians (the Euclides algorithm for finding the greatest common divisor, and the Eratosthenes algorithm for building a sequence of consecutive prime numbers), but the term itself was introduced much later and comes from the Latin form (Algorismi) of the name of a ninth-century mathematician from Baghdad—Muhammad ibn Musa al-Khwarizmi.

What is amazing about modeling is how accurately we can often describe the world, notwithstanding the enormous simplifications we are forced to apply. In this respect, nature is kind to us, even though it could have been altogether different. All phenomena could have been intrinsically intertwined with one another, and without disentangling the entire knot of dependencies we would not be able to draw conclusions regarding the behavior of any single thread. Moreover, it is possible that there exist fields in which no simplified models will yield a suitable description of reality, and in which one must take into account an entire multitude of mutually-dependent factors to obtain a result. One good candidate for such a model-defiant object is our consciousness. In most other areas of research, however, modeling seems to bring success upon success and so far no limits imposed by nature are making themselves apparent. Progress seems to be purely a function of time.

One could come to the conclusion that *every* theory—every description of reality—is a model, since evidently no theory is perfect. What is then the difference between the modeling of a phenomenon and the creation of a theory describing it? The borderline between these two concepts is indeed diffused. Some theories are nothing but simple models, while some models deserve the rank of a fully-fledged theory. In our opinion, however, the difference lies in the fact that, when forming a **theory**, we try to take into account all of the factors we know to have an influence on the studied phenomenon. When building a theory we aim at perfection (even though we rarely achieve it).When constructing a model, however, we *purposely* exclude some of the factors and leave only a few chosen ones for the sake of obtaining a simpler scheme.

The omission of influential factors can be complemented by the use of **probability theory**, a very helpful instrument, to which we devote a big part of this book. It allows us to replace *full knowledge* of the studied phenomenon with what we might call *average knowledge*. In this way we are able to achieve approximate results or the most probable results, which are often the best we can get when full knowledge is practically beyond our reach.

We start by introducing a simple model, which shall serve us as a paradigm of modeling reality throughout this book, be it social, biological, physical, or any other kind of reality. It is the cellular automaton called *The game of life*.

The game of life

A legendary cellular automaton

In the late 1960s John Conway devised a one-person game, which he called *The game of life*. In 1970 Martin Gardner, the author of a mathematical games section in *Scientific American*, popularized Conway's idea. Gardner's two articles caused utter fervor among many computer enthusiasts. Personal computers were not yet around at that time (they appeared only in the mid-1970s) and all experiments with *The game of life* were conducted on giant, expensive mainframes. It has been said that such *Life* games cost American companies several million dollars. The source of such fascination must have been the extreme simplicity of the game's rules combined with the vast variety of resulting 'life-forms' and wealth of interesting problems which arose from it.

John Horton Conway (1937–); English mathematician from Cambridge; known, among others, as coauthor of Conway and Guy (1996).

The game of life represents a very simple model of the birth, evolution, and death of a colony of living organisms—let us call these organisms bacteria. Theoretically, one can play the game by moving around tokens on a *Go* board, although this is a very tedious task. That is, nonetheless, how the game was played in the beginning. To avoid mistakes, two colors of tokens were used. Each move was realized by replacing all of the tokens on the board with tokens of the other color, representing the next generation of bacteria, according to the game's rules described below. Yet it was only the utilization of computers that gave the game its full brilliance.

The rules of the game

Like many models of reality, as well as many board games, *The game of life* takes place in space and time. Space in this game is a two-dimensional lattice divided into square cells (Fig. 2.1). Each cell contains exactly one bacterium—this bacterium can be *dead* or *alive*. A cell containing a living bacterium is marked, while an unmarked cell contains a dead bacterium. The flow of time is also measured discretely, by numbering the consecutive generations of the colony.

The game begins by selecting the cells occupied by the initial generation of living bacteria. The second generation results from the first one as the rules of the game are applied, and so on. In this fashion any number of subsequent generations can be constructed.

Conway experimented for a long time, trying out different rules for the development of the colony. Finally, he came across a set of rules that makes

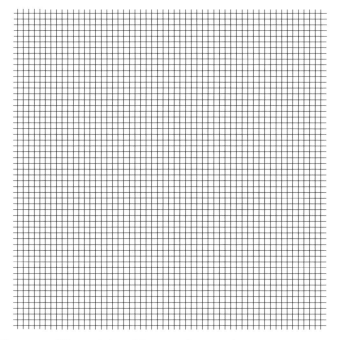

Fig. 2.1 *The game of life* empty board (or computer screen).

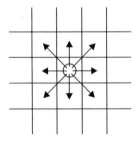

Fig. 2.2 The eight neighbors of a cell.

the evolution of the system both interesting and hard to predict. Conway's rules determine

 (i) when a bacterium survives and passes on to the next generation,
 (ii) when a new bacterium is born in place of a dead one, and
(iii) when a living bacterium dies.

Which of the events is to take place is determined by the number of living neighbors of a given cell. Each cell neighbors with eight others, as depicted in Fig. 2.2. Thus, the rules are as follows.

- A bacterium with zero or one living neighbors dies of solitude.
- A living bacterium with two or three living neighbors is happy and survives on to the next generation.
- A dead bacterium with exactly three living neighbors is reborn due to optimal living conditions.
- A living bacterium with four or more living neighbors dies of over-crowding.

To make it more straightforward, we can abandon the bacteria concept and refer simply to the cells as being dead or alive. Following this, we can formulate Conway's rules in two sentences.

- If a cell is dead, then it is reborn if and only if it has exactly three living neighbors.
- If a cell is alive, then it dies if and only if it has less than two, or more than three, living neighbors.

We leave the demonstration of the equivalence of the above two sets of rules to the reader as a simple exercise in logic.

With each tick of the clock we conduct a full survey of our cells and, according to the rules explained above, we apply the changes by registering all newborn cells and deregistering the ones that died of solitude or over-crowding. Of course, all of these tasks can be done for us by a computer. Figures 2.3 and 2.4 portray the development of a small colony of bacteria

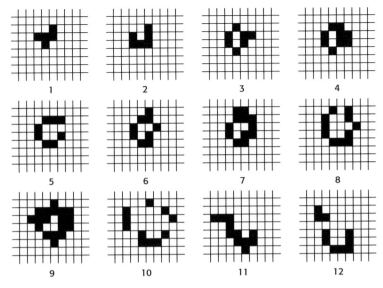

Fig. 2.3 The evolution of a colony composed initially of five living bacteria.

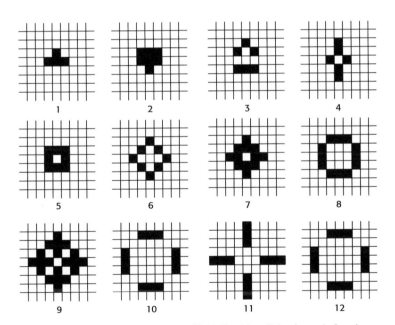

Fig. 2.4 The evolution of a colony composed initially of four living bacteria forming a triangle. After ten generations we get a recurring pattern known as 'the traffic Light'.

throughout twelve generations. We encourage readers to check that each step is constructed from the previous one in accordance with the rules of the game.

The program **Conway** serves to generate a sequence of *Life* generations on a computer. We can use it to track the evolution of a colony constructed by us, or start from one of the pre-built configurations known to yield interesting results. Of the latter, we recommend trying the 'glider factory' configuration—it was devised in answer to one of the problems posed by Conway: is there a configuration whose size (number of living cells) grows without limits?

What we find most amazing when observing the successive generations of bacteria is the giant contrast between the complexity of the development of a colony and the simplicity of both the rules, and the initial configurations. A wealth of forms can therefore be a consequence of very simple rules, and in this respect *The game of life's* case is not an isolated one. More 'scientific' examples can be found in physics or chemistry. Millions, or even billions, of organic chemical compounds possess an immense range of remarkable properties; and yet, they are constructed out of carbon, oxygen, and hydrogen atoms, and a few other elements, in accordance with rules which can be inscribed in the form of one equation—the Schrödinger equation. Finding such simple rules to explain a complex set of phenomena is indeed the quintessence of modeling.

Cellular automata

The game of life is an example of what scientists call a *cellular automaton*. Cellular automata, the theory behind them, and their application are now considered to be a separate field of science. Its beginnings date back to the late 1950s, when Stanislaw Ulam and John von Neumann devised a method for computing the motion of a liquid. The principle of the method was to divide the liquid into cells and determine the motion of each cell on the basis of the behavior of the neighboring ones.

Some consider the Pascal triangle (see Chapter 4) to be the first cellular automaton, with its subsequent rows signifying the successive generations.

Stanislaw Marcin Ulam (1909–1984); a Polish mathematician residing in the United States from 1939. During the Second World War he took part in the Manhattan Project in Los Alamos, where he was the first to present a realistic technique for the construction of the hydrogen bomb.

John von Neumann (1903–1957); an American mathematician of Hungarian origin. Renowned not only as a mathematician, but also as a theoretical physicist and initiator of many fields of applied mathematics. He created game theory (1928), presented the first algorithm for generating random numbers (1946), formulated the functioning principles of a modern computer containing a central processor (CPU) and data storage memory (RAM) (1945), and played a large role in the construction of the first nuclear bomb.

In general, a cellular automaton can be characterized by the following three parameters:

- the structure of the space in which all events take place;
- the definition of objects inhabiting that space;
- a set of rules defining the behavior of these objects.

Due to the huge diversity of conceivable cellular automata, a uniform theory of such structures has not yet been developed. Only in the simplest case of a one-dimensional space and with the introduction of other limiting assumptions did Stephen Wolfram arrive at a complete classification of cellular automata. If we decided to allow each cell of an automaton to have one of only two possible values (like in *The game of life*) and if the evolution of a cell depended only on its own value and on the value of two of its closest neighbors, then we would already have 256 possible automata. This follows from the fact that there are $2^3 = 8$ possible states of three neighboring cells, and for each such state we can independently select one of two possible outcomes—the resulting state of the middle cell. We get $2^8 = 256$ possibilities. The number of possible automata grows dramatically with the increase in the number k of possible cell states and the number s of cells influencing the evolution of a given cell. The general equation is k^{k^s}. If we increase the range of influence to include two other neighbors (now $s = 5$, i.e. the cell itself and four of its neighbors), then we already have $2^{2^5} = 2^{32} = 4\,924\,967\,296$ possible different cellular automata.

Stephen Wolfram (1959–); creator of Mathematica® and founder of Wolfram Research (http://www.wolfram.com). His recently published book (Wolfram, 2002) is a giant tribute to cellular automata, which the author claims underlie all of the universe's phenomena.

It is worth noting that the future of a cellular automaton can always be predicted, as long as its rules do not contain any elements of chance. Usually, however, knowing the state of an automaton does not allow us to guess

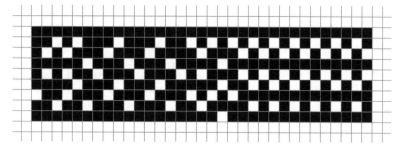

Fig. 2.5 The first 'Garden of Eden'-type configuration, found by Roger Banks—a configuration for which no previous generation exists.

its past. A configuration can have many possible predecessors, or it can be a configuration for which no previous configuration exists. A configuration with no possible predecessors has been given the evocative name of a 'Garden of Eden'. The quest for Gardens of Eden was a popular activity during the beginnings of *The game of life* craze. Roger Banks was the first to find such a configuration (depicted in Fig. 2.5) in 1971, with the use of a computer program searching all possible preceding configurations.

Cellular automata have found many applications in physics, chemistry, and technical sciences. We can use them to model such diverse processes as the motion of loose matter (such as sand piles), the flow of fluid through porous materials, the spread of forest fires, traffic jams, the interaction of elementary particles, and many others. Also, the models of society described in Chapter 14 are very much like cellular automata. Neural networks, which model the human brain and are described in Chapter 13, have a lot in common with cellular automata as well, since each element of such a network functions, to a large degree, independently.

3

Heads or tails

Probability of an event

The game of life described in the previous chapter is fully **deterministic**—its rules do not contain any elements of chance. Each generation of bacteria *uniquely* determines the next generation. Predicting the future is possible in this case, since at each stage we possess the full information regarding the state of the colony. We know which bacteria are alive, which are dead, and exactly which rules shall be applied to produce the next generation. However, with more complex models of reality this is often not the case as our knowledge of the studied system is frequently incomplete. The description of such models requires utilizing the notion of **probability**. The field of science dealing with this concept is called **probability theory**.

The ancient Greeks divided phenomena into two groups: those which conformed to the laws of nature, and those which were completely independent of these laws—they were governed by chance. Nowadays, we know that many seemingly random phenomena have a deterministic origin. This information, however, is often of no practical use to us—when the amount of information needed to describe the system and to predict its behavior is too large to process. Let us consider, for example, the problem of predicting the behavior of a liter of gas. The gas is composed of molecules whose behavior conforms to the laws of physics. If we know the position in space of one such molecule, its speed, and the direction in which it is moving, then we can pretty well predict where it can be found in the future. The problem is that a liter of gas contains about 10^{23} molecules. We need six numbers to describe the state of one molecule (three coordinates for the position and three coordinates for the velocity). So, even if we were to give each person on Earth a computer with a 100-gigabyte hard disk, then we would still not come close to storing the information needed, much less would we be able to process it in any way and in any reasonable amount of time.

Fortunately, we also know that the whole set of molecules conforms to laws of statistics—they are *more or less* evenly distributed in the space that contains them, and they move *more or less* the same amount in every direction and with some *average* speed. The intelligent use of this information in addition to probability theory allows us to model very accurately the behavior of the whole liter of gas without ever having to model the motion of each single molecule. Owing to the simplifications brought about by the use of statistical theory, we are able to predict the (average) behavior of the gas in realistic time with the use of a regular computer (or perhaps even just a pen and paper).

The need for a statistical description, one which utilizes average values, has an additional justification. The contemporary physical theory of the world—quantum mechanics—assumes that the behavior of micro-objects (such as electrons, photons, atoms, or molecules) is governed by laws, which can *only* be formulated with the use of probability theory. Moreover, the state of such objects (for example, their position and velocity) can never be determined exactly but only with a certain probability. Thus the laws governing macroscopic objects are deterministic (and only on average) because they deal with values averaged over an enormous number of micro-objects. So, the ancient Greeks were, in a way, right. The impossibility of exactly determining the actual state of things means that in reality many phenomena are indeed governed by chance. The difference between the ancient and contemporary state of knowledge is that we learned, taking advantage of probability theory, to utilize our incomplete knowledge and draw precise, testable conclusions.

What are the chances?

We will begin our introduction to probability theory with the simplest of all examples, namely tossing a coin. Assuming that the coin is perfectly balanced, neither of the two possible results is favored. Of course, real coins are never ideally symmetric, but common sense suggests that the influence of this asymmetry is small enough to be insignificant. In other words, we can assume that it is equally probable to obtain a head as a tail. In the language of probability theory, we would say that the probability p_h of getting a head equals the probability p_t of getting a tail. It has been agreed that the sum of the probabilities of all possible outcomes of a trial is equal to one. This is indeed only a convention. For example, when working with percentages, we would use a different convention and that sum would be 100. When tossing a coin, the only possible outcomes are head and tail. Thus the probability of each of these results, p_h and p_t, must be equal to 1/2.

What is **probability** in general and how can we define it?

> Probability is a number between zero and one, assigned to a potential outcome of a trial. This number is a measure of the chance that the trial will have that particular outcome.

In the simple case of tossing a coin, we managed to guess the probabilities of the two possible outcomes based on reasoning about **symmetry**. In the case of tossing a symmetrical coin, none of the two possible outcomes is distinguished. Similar cases are those of rolling an unbiased dice, or picking a card out of a shuffled deck. Rolling a dice has six equally probable outcomes, since none of the sides is distinguished and so each of the results has a probability equal to 1/6. Reasoning along the same lines, we can conclude that the probability of picking a specific card from a shuffed deck is equal to 1/52.

Another interesting example of the lack of distinction leading to equally probable results is the decimal representation of some irrational numbers. Of course, not all numbers have a decimal representation in which none of the digits are distinguished. Rational numbers do not have this property since they are periodic, which means that their decimal representation always has the property that from a certain point it starts repeating itself over and over. For example, the decimal representation of 1/6 is 0.166666..., where the sequence of sixes is infinite. Also, not all irrational numbers have this property. For example, one could take a number with a decimal representation which follows the pattern 1.211211121111211111..., and so on. It is not rational since it does not have a period and yet it is composed of mostly ones, some twos, and no other digits. However, the examination of such irrational numbers as π, $\sqrt{2}$, or e (the base of natural logarithms) shows that none of the digits seem to be distinguished. This observation is confirmed by Table 3.1, which shows that each digit occurs approximately 100 000 times in a million-digit decimal representation of each of these numbers.

We generated the decimal representations of the numbers with Mathematica® and calculated the digit statistics with our program **Poe**. By the term statistics we mean here the collection of data representing the distribution of random elements rather than the branch of mathematics called

Table 3.1 The number of occurrences of each digit in the million-digit decimal representations of the irrational numbers π, $\sqrt{2}$, and e.

	π	$\sqrt{2}$	e
0	99959	99814	99425
1	99757	98925	100132
2	100026	100436	99846
3	100230	100190	100228
4	100230	100024	100389
5	100359	100155	100087
6	99548	99886	100479
7	99800	100008	99910
8	99985	100441	99812
9	100106	100121	99692

statistics. In this chapter we will deal only with the simplest statistical properties of random events—that is, average values and departure from the average. The next chapter deals with more complex statistical properties.

Among other features, the program **Poe** calculates how many times each character, pair of characters, or triplet of characters occurs in a given sequence. This program will be described in detail in Chapter 9.

If we take the single-digit occurrences and pair occurrences calculated by Poe and divide them by the total number of characters in the sequence, then we get the probabilities of the occurrence of each digit and each pair. In one of the million-digit representations the average number of occurrences of a digit is 100 000 and of a pair is 10 000, since there are 10 different digits and 100 different pairs. The actual results differ from the average by no more than 0.5% for single characters. For pairs, the difference between the theoretical value and the actual outcome reaches 2%, since the number of occurrences of each pair is about ten times smaller than that of a single digit and therefore the difference is greater. (For a single flip of a coin the departure from the mean value is always 50%!)

An even distribution of characters does not characterize just the decimal representations of these numbers. Representations in any other base have the same property. For example, the binary representations are random sequences of zeros and ones, even though this randomness is apparent only after examining a longer segment of the sequence. The beginning of the binary representation of π is evidently dominated by zeros (124 in the first 204 characters, which is a departure of almost 22% from the average value). The difference diminishes (though slowly) with the length of the examined sequence, and after 26 596 characters the number of ones catches up with the number of zeros. This example serves to show that large **fluctuations** can be found even in undoubtedly random sequences.

> We would like to turn the reader's attention to the enormous progress taking place in the field of computer calculations. Calculating the number π with million-digit accuracy took less than a minute on our PC. In 1949, generating 2037 digits of π took around seventy hours on the first real computer—ENIAC (Electronic Numerator, Integrator, Analyzer, and Computer, a thirty-ton, 17 468-lamp, and 174-kilowatt machine). In 1962, the computer scientists Daniel Shanks and John W. Wrench published an article ('Mathematics of computation') in which they presented a record-breaking result—a table of the first 100 000 digits of π. Their calculations on the powerful (at that time) IBM 7090 took almost nine hours. Nowadays, powerful computers can effortlessly generate a billion digits of π.

Theory and practice

The theoretical value of probability, arrived at with the help of arguments about symmetry—a value given a priori—can be contrasted with the result of experiments—a value obtained a posteriori. Such experiments have been conducted for hundreds of years. In the eighteenth century, George-Louis Buffon conducted the first documented experiment of this sort by tossing a coin 4040 times. He got 2048 heads and 1992 tails, which results in an average value of 0.5069. The average value is defined by assigning the value zero to a head and one to a tail result. Ideally, the average value should, of course, be half-way between zero and one, and hence be equal to 1/2. A record was set in the twentieth century by a Russian statistician Vsievolod Ivanovich Romanowski, who tossed a coin 80 640 times, obtaining an average of 0.4923. It is easy to estimate that this task must have taken him more than twenty-four hours. Several other examples of such experiments can be found in Grinstead and Snell, 1997.

Count George-Louis Leclerc Buffon (1707–1788); French scientist and writer.

Richard von Mises (1883–1953); Austrian mathematician and engineer born in Lvov; professor at the University of Berlin. When Hitler came to power, he emigrated to Turkey and then to the United States.

Studying such trends is much easier nowadays, since we can be aided by computers. An average computer can easily simulate 100 000 coin tosses in much less than a second. We leave until Chapter 4 the discussion on whether one may trust a computer to generate random numbers and what the limitations of this process are.

The program **Buffon** can be used to simulate the tossing of a coin a given number of times. It calculates the number of ones obtained in such a trial, as well as the deviation from the expected average value. It also shows how the percentage of ones changes throughout the experiment.

By running **Buffon**, one can see that the more times a coin is tossed, the closer the percentage of ones gets to 50%. This is in perfect accordance with our intuition. It is also the basis of the so-called **limiting frequency theory** proposed by Richard von Mises, which states that the probability p_A of an outcome A is the limit

$$p_A = \lim_{N \to \infty} \frac{N_A}{N}.$$

In the above formula, N_A denotes the number of trials with outcome A and N is the overall number of trials. When using this definition, one should not, of course, treat the limit in its strict mathematical sense, since we can never perform an infinite number of trials. It simply means 'a very large number of trials'. Mathematicians are very skeptical about the von Mises definition, arguing that it is not precise. The frequency definition can, however, be useful when applied to problems which do not contain any elements of symmetry. For example, demographic studies conducted for thousands of years (the first demographic studies were conducted in China over 2000 years BC) showed that the ratio of male births to all human births is constant and equal to about $22/43 \approx 0.5116$. The probability of a male birth was considered to be such a well-founded fact that Pierre-Simon Laplace knew something extraordinary must have happened when this ratio for Paris in the years 1745–1784 turned out to be a little lower, namely $25/49 \approx 0.5102$. Indeed, a subsequent analysis of the town's records showed that the 'birth rate' of girls rose due to the fact that the poorer inhabitants of the Paris vicinity, who tended to drop their babies off in the city, were more inclined to leave their baby girls rather than their boys.

 Pierre-Simon Laplace (1749–1827); French mathematician, physicist, and astronomer; one of the founders of modern probability theory.

The frequency definition is criticized by mathematicians for being inaccurate, yet it is undoubtedly very important to empirical sciences, in which experimental data is the only source of knowledge. One would, of course, be inclined to adopt the following definition proposed by Laplace.

The probability P_A is equal to the ratio of the number of favorable cases to the overall number of possible cases.

The problem, however, is determining the 'number of favorable cases' when one does not possess a complete knowledge concerning the studied events and cannot, for instance, utilize symmetry. For example, one cannot use Laplace's definition (with the modern state of genetics) to obtain the magic number 22/43. The theory of probability developed over the years by mathematicians is an exact branch of mathematics, yet it leaves unanswered the question: how can one, in specific cases, find the probabilities of all studied events, or the **probability distribution**? One can only achieve this by trail and error— guess the distribution, and if observations agree with it then the distribution has been guessed correctly.

The definition of probability based on favorable cases is filled with traps awaiting the inexperienced. The problem of calculating the number of possible cases does not always have a simple solution. Let us consider a very simple example of tossing two die and let us ask what the probability is of obtaining an even sum of the numbers on both die. There are several ways of finding the solution to this simple question. We can, for example, utilize what we know about the probability distribution when tossing a coin, and identify obtaining an even number on one dice with getting a head and an odd result with getting a tail. In this case, getting an even sum on two die would be equivalent to getting head–head or tail–tail in two tosses, and an odd result would be equivalent to a head–tail or tail–head result. This would imply that the probability of obtaining an even sum is equal to the probability of getting an odd sum and that both are equal to 1/2, which does not seem surprising. We can, however, count the number of favorable events in another way. We can enumerate all of the possible sums obtained when throwing two die and then count the percentage of even ones. The possible sums are 2, 3, 4, 5, 6, 7, 8, 9, 10, 11, and 12. The majority of these numbers are even. Applying the definition blindly would lead us to believing that the probability of obtaining an even sum on two die is equal to 6/11. This result does not agree with what we argued before and hopefully also not with our intuition. Where then is the mistake? In the enumeration of cases! The cases 2 and 12 are actually proper cases (so-called **elementary events**), but the rest are in fact bunches of several possible events. For example, the event 7 is actually a combination of the six different elementary events $1 + 6$, $2 + 5$, $3 + 4$, $4 + 3$, $5 + 2$, and $6 + 1$. If we count each such elementary event separately, then we will find that there are $6 \times 6 = 36$ of them and exactly eighteen yield an even sum, so the probability is indeed 1/2. In the case of throwing two die, the correct enumeration of cases and the identification of favorable ones was easy; however, in more complex situations we might find that this task is much harder to accomplish.

The continuous case

Laplace's definition has yet another shortcoming which greatly troubled mathematicians: it cannot be applied when the number of possible cases is infinite. There are times when such situations can in fact be correctly described, but there are others which yield unexpected paradoxes. A very simple example of a problem with an infinite probability space is the following. What is the probability that when entering a subway station we will find the train there, if we know that trains run every ten minutes and spend one minute at the station. Even though the moment of our arrival at the station is an element of an infinite set of such possible moments, we have no doubt that the probability is 1/10. The obviousness of this solution comes from the existence of a natural measure—the measure of time. We can divide the entire span of time into segments and classify those segments as favorable (the train is at the station) and unfavorable (the train is not at the station). The probability of finding the train at the station is the ratio of the total length of the favorable segments to the length of the entire considered time-span, and hence 1/10.

In order to solve the above problem, we were able to substitute the *number* of cases with a *measure* defining the amount (not necessarily finite) of cases. The solution was simple, because the choice of the measure was obvious. This, however, is not always the case, as illustrated by the so-called **Bertrand paradox**. The paradox occurs when seeking the answer to the question: what is the probability that a randomly chosen chord of a given circle will be longer than the side of an equilateral triangle inscribed in the circle? We will describe several different ways to approach this problem. They are all illustrated by the program **Bertrand**.

Joseph Louis François Bertrand (1822–1900); French mathematician and also an economist. In the book Calcul de probabilitess he pointed out the need for a rigorous definition of continuous probabilities.

(i) The first approach takes advantage of the rotational symmetry of the circle. Since the result should be the same regardless of the orientation of the circle, let us rotate it so that the chord C is vertical (parallel to the Y-axis) and inscribe the triangle so that one of its sides is also vertical. Now we can identify all of the cases with the possible X positions of the chord. From Fig. 3.1, we can clearly see that if the chord lies to the left of the point a then it is shorter than the side of the triangle. Analogously, it is also shorter when it is to the right of the point b. (The point b is a reflection of the point a with respect to the vertical symmetry axis of the circle.) In all other cases it is longer. Simple geometric calculations show that the distance between a and b is one-half of the diameter of the circle, which would lead us to believe that the solution to the problem is $1/2$.

(ii) For the second approach, let us inscribe the triangle so that its one corner is at one end of the chord. Now, all of the cases are covered by the possible angles formed by the chord and the diameter of the circle, as shown

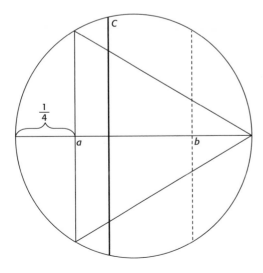

Fig. 3.1 We orient the circle so that the chord C is vertical and consider all of its possible positions along the X-axis.

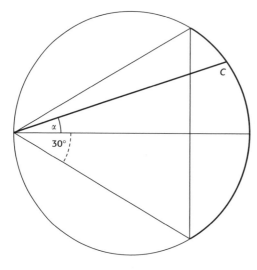

Fig. 3.2 We fix one end of the chord C and consider all of the possible positions of the other end (or all possible angles).

in Fig. 3.2. This angle is always between $-90°$ and $90°$. If it is between $-30°$ and $30°$ then the chord is longer than the side of the triangle, otherwise it is shorter. Therefore, 60 out of the 180 degrees satisfy the condition, so the probability should be 1/3. Instead of considering the angles, we could consider the positions of the second end of the chord as the possible cases. For the chord to be longer than the side of the triangle, its second end must lie on the smaller arc of the circle between the other two corners of the triangle, i.e. the thicker curve in Fig. 3.2. This arc is one-third of the circumference of the circle, so the probability is again 1/3.

(iii) For our third approach, let us identify all of the possible cases using the positions of the center of the chord. Each point within the circle is the center of exactly one chord. Some elementary geometry shows that only the chords with a center inside the circle inscribed in the inscribed triangle are longer than the side of the triangle (see Fig. 3.3). The radius of the inscribed circle is one-half of the radius of the large circle, so the ratio of their areas is 1/4. This would mean that the sought probability is 1/4.

(iv) There exist numerous other ways to approach this problem. For example, one could pick two different random points inside the circle and calculate the probability that the chord passing through these points is longer than the inscribed triangle's side. After some tedious calculations, this approach gives the result $1/3 + 3\sqrt{3}/4\pi \approx 0.74683$. Perhaps you can find yet another way to solve this problem?

Which of these solutions is actually correct? All of them are! The point is that we have not specified from the start what a random chord is. In each of

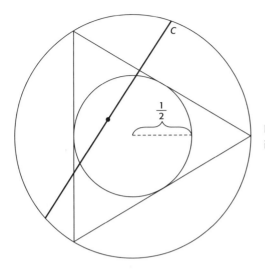

Fig. 3.3 We identify a chord by its center point.

the approaches we assumed a different interpretation of 'a randomly chosen chord', and hence got different results. Discrepancies such as this one do not appear when all of the cases can be numbered, as is the situation when tossing a coin or throwing a dice. There is also no problem when we can assume an even distribution of the randomly chosen quantity, as occurred in the train station problem. However, when dealing with several quantities characterizing an event (in the case of the chord these were the coordinates of points and the angles) saying 'randomly chosen' is usually ambiguous.

Shooting blindly

Finally, we would like to show how one can use random numbers to perform calculations. This technique is called the **Monte Carlo** method, since it has a lot in common with a casino, in which chance plays a decisive role. The method was invented by Stanislaw Ulam; when recovering from a serious illness he took to playing the solitaire game Canfield. While trying to calculate the probability of winning, Ulam came to the conclusion that it would be much easier to play a hundred times and use the outcomes of the games to approximate the chance of winning, rather than to calculate the exact result by means of rather complex combinatorial calculations. This simple yet brilliant observation is the basic idea of the Monte Carlo method.

A simple application of this method is in the calculation of area under the graph of a function $f(x)$ over the range (a, b); in other words, the calculation of the **integral** of $f(x)$ from a to b. In order to achieve this, we generate some random points within the given range, calculate the average value of f at these

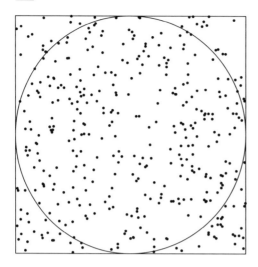

Fig. 3.4 Calculating π using the Monte Carlo method. Here 315 out of 400 points lay inside the circle. This gives the approximate value $\pi \approx 4 \cdot 315/400 = 3.15$.

points, and multiply it by the length of the range. This method is not the best for calculating integrals of one-dimensional functions; however, it does start to pay off with multiple integrals—when the function depends on more than one variable i.e., $f(x_1, x_2, \ldots, x_n)$, and we need to calculate the volume of an $(n + 1)$-dimensional solid under the curve.

The Monte Carlo method can also be used to approximate the value of π. If we inscribe a circle of radius 1 into a square of side 2, then the ratio of the area of the circle to the area of the square is $\pi/4$. We can therefore generate random points from within the square and note which of them fall within the circle (see Fig. 3.4). As we keep performing such trials, the ratio of the number of points inside the circle to the total number of trials should approach $\pi/4$. This example serves solely to illustrate the principle of the method, since there exist far better algorithms for determining π.

Another example of how the Monte Carlo method leads quickly to an approximate solution is our last approach to the Bertrand problem presented earlier in this chapter. Arriving at the exact solution required long and tedious calculations, while finding the approximate result by writing a program based on the Monte Carlo method took just a few minutes.

The program **Ulam** illustrates how the Monte Carlo method can be used to calculate integrals of a given function within a given range.

4

Galton's board

Probability and statistics

Experience shows that when tossing a coin we will not get a sequence of perfectly alternating heads and tails. It is true that such a result would not be in conflict with the probability distribution, but there is more to it. All of us must have noticed that some *additional* regularities appear when tossing a coin—regularities which go beyond the mere probability distribution. Before we discuss this phenomenon in detail, we would like the reader to take part in the following simple experiment.

- Launch the program **Bernoulli** and manually type in a sequence of 0s and 1s. Try to make the sequence as random as you can—the way you would imagine the result of subsequent coin tosses would look—but do not actually toss a coin, just invent the sequence. Make the sequence 200 or more characters long.
- Press 'Generate statistics' to calculate the occurrences of continuous blocks of 1s different length in your sequence.
- Without closing the chart window, switch to automatic mode and generate a random sequence of the same length as the one you typed in, using the built-in random number generator.
- Press 'Generate statistics' to calculate the occurrences of continuous blocks of 1s of different length in the computer-generated random sequence.
- Compare the two charts. Do you see a difference?

The program **Bernoulli** calculates occurrences of continuous blocks of 1s in a manually-entered or computer-generated random sequence.

If you do not have access to a computer then you can perform the above experiment on a piece of paper. Write the sequence down and, once you have finished, locate all of the continuous blocks of 1s in it. Count how many of these blocks are of length one, length two, etc.

 Jacob Bernoulli (1654–1705); Swiss mathematician. His work '*Ars conjectandi*' ('The art of conjecturing') became the foundation of modern probability theory, containing, among other things, the following definition: '*Probabilitas enim est gradus certitudinis, & ab hac differt ut pars à toto.*' ('Probability is a degree of certainty, and differs from certainty as a part from a whole.')

Naturally, we have no way of knowing what sequence the reader invented, but our guess is that the two charts differ significantly. It turns out that the vast majority of people, when producing 'random' sequences of digits, tend to avoid long sequences of the same digit. The computer-generated sequence, however, will most likely contain some blocks of six or even more 1s in a row.

Is there something wrong with our intuition about random numbers or are random number generators biased in some way? To answer this question let us calculate how many k-length blocks of 1s should be contained (on average) in a sequence of length n. The exact calculation of this value is very complex, but we can estimate it very accurately by assuming that n is large and that k is much smaller than n. With this assumption, we can disregard the effect that the occurrence of one block has on the number of occurrences of other blocks within the same sequence. We can also disregard the special case of a block occurring at the end or the beginning of the sequence. The calculation is based on the observation that a block of length k occurs in our sequence when a 0 is followed by k 1s and another 0. The 0s at the edges are important, since if one of them was a 1 then the block's length would not be k, but something more. Since the probability of each character is 1/2 and the characters are generated independently of each other, the chance of obtaining a particular sequence of length $k + 2$ (in this case the series $011\ldots110$) is $(1/2)^{k+2}$. The block can appear at (approximately) any of the n positions in the sequence, so the average number of k-length blocks in an n-length sequence is about $n/2^{k+2}$.

Let us compare this result with that obtained from our random number generator. According to our calculation, a 100 000-character-long sequence should contain about 12 500 single 1s, 6 250 doublets, 3 125 triplets, 1 562 quadruplets, and so on. The chances are that it will even contain a block of fifteen or more 1s in a row. Indeed, using **Bernoulli** to generate a few 100 000-character-long sequences and their associated graphs confirms the result of our calculation. Of course, a fair amount of fluctuations can be noted, especially for longer blocks, since they occur less frequently and thus the number of their occurrences has a smaller chance of being averaged out.

It was Alphonse Chapanis who first described the human erroneous intuition about randomness in his 1953 paper published in the *American Psychologist* entitled 'Random-number guessing behavior'. He asked the subjects of his experiments to write long random sequences of decimal digits and observed that most people tend to avoid repeating the same digit three times in a row, even though such triplets occur relatively often in truly random sequences. He also noted that many people consider certain digits to be more 'random' than others.

Alphonse Chapanis (1917–2002); founder of ergonomics (human factors in engineering). As a specialist in this field, he selected the currently-used 3×4 key arrangement on the key-pad for push-button telephones.

Misconceptions of randomness

This little experiment showed that our intuition about random numbers can often mislead us. Another example of this phenomenon is the famous birthday paradox. The probability that a group of thirty people contains two people born on the same day of the year is about 70%, though most people would guess it to be much lower. People who design lotteries often take advantage of our inability to properly estimate a probability. Did you ever buy a lottery ticket and think, 'It was so close! Next time I will surely win.'? If so, then it was most probably an illusion, shrewdly devised by the organizers to induce you to buy more tickets.

The study of human misconceptions of randomness has a very important application, namely it can be used to detect fraud. Since data fabricated by people has statistical qualities which significantly differ from those of real data, one can use a computer program to automatically spot fakes. Of course, such statistics cannot be held as *proof* of fraud, but they can be very useful in identifying suspects. An example of an institution which utilizes this knowledge is the Internal Revenue Service (IRS) of the United States, along with other tax agencies worldwide. The IRS has the means to verify only a small sample of the tax returns it receives. By picking out this sample with the help of statistical tests, they significantly increase the number of fabricated tax returns in the sample. Such methods are also used to detect fabricated surveys, fake scientific data, dishonest bookkeeping, and many others.

The phenomenon which plays the most important role in detecting fraud is Benford's law. The law is stated as follows. Suppose that we pick a number from some set of *natural data*. This set of data could be populations of countries, areas of rivers, results of some scientific experiment, sports statistics, or numbers appearing on a tax return. According to Benford's law, the probability that the first (most significant) digit of this number is 1 is about 30%. The probability decreases logarithmically with each further digit, according to the following formula:

$$P(d) = \log_{10}\left(1 + \frac{1}{d}\right), \quad \text{where } d = 1, 2, \ldots, 9.$$

This law seems to contradict all that we have learned about probability! In fact, even some mathematicians find it hard to believe at first. Since we pick a completely random number, from some completely random source, no numbers should be distinguished, and so no significant digits should be distinguished and all of them should be equally probable. Yet empirical observations show that this law holds for most datasets appearing in everyday life. Moreover, a closer look shows that, paradoxically, it is symmetry, or rather invariance, that lies at the base of this phenomenon.

Benford's law was first observed by the American astronomer Simon Newcomb in 1881, who noticed that books of logarithms were dirtiest at the beginning and progressively cleaner throughout. (Books of logarithms were a common tool for finding logarithms of numbers when calculators and computers were yet to be invented.) At that time, the discovery did not attract much interest and its original author was forgotten. The law's name comes from the General Electric Company physicist Frank Benford, who made exactly the same observation in 1938 and backed it with 20 229 data elements collected over several years, ranging from atomic weights, through drainage areas of rivers, to baseball statistics and numbers appearing in the *Reader's Digest*.

Only recently (in 1996), Theodore P. Hill of the Georgia Institute of Technology published a series of papers containing a thorough mathematical explanation of the phenomenon.

The program **Benford** calculates significant-digit occurrences of various data-sets, including the set of file sizes on your hard disk, and compares it with Benford's distribution.

Running **Benford** should be enough to convince anyone that the law holds; however, it gives us no clue as to why it is so. Even though Benford's law has been known for over a century, its thorough mathematical justification has only been supplied recently. Naturally, Benford's law does not apply to sets of numbers selected from a given range with equal probability. However, meaningful data is almost never like this—the values it contains depend on different factors and are almost never equally likely to appear, even though their particular probabilities are often very hard to establish. For example, a country's population depends on its area, environment, economy, politics, and many other factors. Hence it is obvious that different populations have different probabilities, though these probabilities can at best be guessed based on empirical data.

Although the proper mathematical justification of the first-digit phenomenon is very complex, it is easy to show why the hypothetical *union* of all natural datasets has this property. Let us suppose that there exists a distribution of significant digits which characterizes this union. Then this distribution should not change if the data is scaled, since different sets are measured in different units. For example, some might measure a country's area in square kilometers and some in square miles, stock quotes (which also fit Benford's law very well) are listed in different currencies, and so on. In other words, such a distribution would have to be *scale invariant*. Moreover, the distribution should not depend on the base that the numbers are written in. After all, the decimal base is a purely human invention and does not have any other natural justification (except maybe the fact that humans have ten fingers). So the distribution should also be *base invariant*. It turns out that base invariance follows from scale invariance, and the only distribution that is base and scale invariant is the logarithmic distribution. This reasoning shows that an even distribution of significant digits would not be 'symmetrical' at all, since changing the units of measure could yield a completely different distribution. The only truly symmetrical distribution is the one given by Benford's law.

There is also a generalized version of Benford's law and it states that, if the set of natural data is written in base b, then the probability that the first digits of a randomly chosen number from this set form the number n (disregarding the decimal point, leading zeros, and the minus sign) is equal to

$$\log_b \left(1 + \frac{1}{n} \right).$$

For example, the probability that the first, second, and third digits of a number in base 10 are 1, 2, and 3, respectively, is $\log_{10}(1 + 1/123)$.

Benford's law has been successfully applied to detect various kinds of fraud, since most people are not aware of its existence and erroneously assume that, in order to make data seem 'believable' all digits should appear just as often. The law can also be used to discover errors in data, whether they are of human or machine origin. Of course, it will not help to track one error in a large database, but it can detect the effects of an erroneous computer program or a faulty clerical procedure. Benford's law can also be used to test computer and mathematical models. If our 'model of reality' is designed to mimic a natural phenomenon, then the data which it is supposed to model is very likely to follow Benford's law. Checking whether the data produced by the model conforms to Benford's distribution is very easy and provides immediate feedback on the adequacy of the model.

> The idea of using Benford's law to detect fraud was put forward in 1992 by Mark J. Nigrini in his Ph.D. thesis entitled 'The detection of income evasion through an analysis of digital distributions'. Since then, it has had numerous successes in detecting fraud worldwide.

Random walking

Let us return to the results of multiple coin tosses and ask the following question: what numbers are we likely to get after n tosses if, starting from zero, we add one in the event of obtaining a head and subtract one in the event of obtaining a tail? This problem can be pictorially modeled by the **Galton board**—a slanted wooden panel covered with a triangular pattern of nails (see Fig. 4.1). Metal balls rolling from the top of the board encounter subsequent nails and bounce left or right with equal probability. After having traversed the pattern of nails in such a manner, they fall into one of the slots below. Each bounce of a ball can be identified with a toss of a coin and the resulting slot with the resulting sum. If the ball bounces left every time, then it will land in the leftmost slot. Similarly, if the tossing each time yields a tail, then the sum will be $-n$. By symmetry, the rightmost slot corresponds to obtaining n heads. It is easy to calculate that the probability of reaching one of these outer slots is equal to $1/2^n$, since only one path out of the 2^n possible paths corresponds

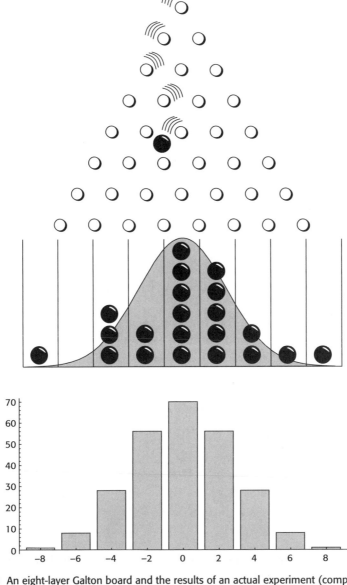

Fig. 4.1 An eight-layer Galton board and the results of an actual experiment (computer simulated). The continuous curve depicts the Gaussian distribution obtained at the limit, when the number of trials and pegs reaches infinity and the diameters of the pegs and ball, as well as the distance between the pegs, reaches zero. The bar chart below represents the number of paths leading to each slot.

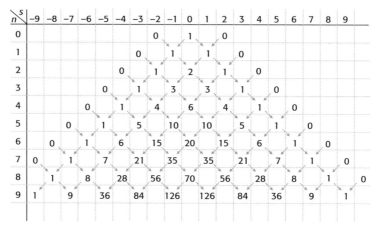

Fig. 4.2 The construction of the Pascal triangle—the number of paths leading to each slot in an n-level Galton board for different values of n.

to each of these outcomes. The number of paths leading to the other slots is a bit harder to calculate.

Notice that getting a sum of s in n tries can be achieved in exactly two ways, namely getting $s-1$ in $n-1$ tries and then one head (or bounce right), or getting $s+1$ in $n-1$ tries and then one tail (or bounce left). Obviously, if $n = 0$ then we have one possible outcome ($s = 0$). We can use these observations to calculate the number of paths leading to each slot, first for $n = 1$, then for $n = 2$, $n = 3$, etc. We start by placing a 1 corresponding to $n = 0$, $s = 0$ and then fill in each of the other values by adding the two values directly above it (see Fig. 4.2).

Francis Galton (1822–1911); English scientist; the first to apply statistical methods to anthropology.

The values obtained by following these rules make up the **Pascal triangle**. The nth row represents the number of paths leading to each slot in an n-level Galton board. It also lists the coefficients in the expanded form of the polynomial $(x + 1)^n$—the so-called binomial coefficients. The binomial coefficients are denoted by $\binom{n}{k}$, where k ranges from 0 to n (and not from $-n$ to n, as in the slot interpretation). The following formula holds:

$$\binom{n}{k} = \frac{n!}{k!(n-k)!}, \quad k = \{0, 1, 2, \ldots, n\}.$$

Here $\binom{n}{k}$ is also the number of ways to choose exactly k objects out of n different objects, disregarding order. This agrees with our model, since we are choosing the positions in the path at which the ball will bounce to the right out of the n bounces making up the path—at the remaining positions the ball will bounce to the left. There is one way to choose zero positions, and hence one path leads to the leftmost slot. There are n ways to choose exactly one right bounce, and so n paths lead to the second slot, and so on.

The Pascal triangle is named after the French philosopher Blaise Pascal (1623–1662) who wrote an important treatise on the subject, though he was not the first to discover it. It is possible that the triangle was known in India as early as 400 BC and it is described in Chinese books dating back to the thirteenth century. The Italians call it the Tartaglia triangle after their famous mathematician Niccolò Tartaglia (1499–1557).

Since the total number of paths is 2^n, the nth row in the Pascal triangle should add up to 2^n; and indeed it does. Therefore, if we divide each value by 2^n then we will get the probabilities of reaching each slot. For example, for $n = 8$ they will be

$$\{p_k\} = \left\{ \frac{1}{256}, \frac{8}{256}, \frac{28}{256}, \frac{56}{256}, \frac{70}{256}, \frac{56}{256}, \frac{28}{256}, \frac{8}{256}, \frac{1}{256} \right\}.$$

 The program **Galton** simulates the behavior of a 'real' Galton board and compares the results with a graph of the Gaussian curve, which will be described in the following paragraphs.

There is yet another way to interpret the results obtained by means of the Galton board, and it is as follows. We toss a coin and, depending on the result obtained, we take either one step left or one step right. Continuing in this manner, we will follow some random route from the starting-point. This process is called a one-dimensional **random walk**. It can be generalized to any number of dimensions. In two dimensions one would take each step north, south, west, or east with equal probability. The third dimension adds two more options, namely up and down. Similarly, each subsequent dimension adds two extra possibilities—moving by one unit forward or backward in this dimension.

The random walk models a phenomenon very common in Nature—**Brownian motion**. It is a pattern which characterizes the movement of small particles suspended in a liquid. In the case of many such particles, the process is called diffusion. The particles move around erratically within the liquid and eventually come to occupy the entire space in a uniform fashion. The theory behind the Galton board and the random walk can therefore be very useful in modeling physical processes of such a nature.

The results obtained with the help of the Galton board are the probabilities of reaching a given point after n steps of a one-dimensional random walk. For example, the probability of returning to the starting-point after n steps is equal to

$$p(n) = \frac{n!}{2^n (n/2)!(n/2)!}.$$

Note that the above equation holds only for even n. For odd n this probability is equal to 0, since there is no way that an odd number of steps can lead back to the starting-point. For $n = 2, 4, 6, 8, 10$, and 12 these probabilities are equal to $1/2$, $3/8$, $5/16$, $35/128$, $63/256$ and $231/1024$, respectively. The general formula for the probability can be approximated by

$$p(n) \approx \frac{\sqrt{2}}{\sqrt{\pi n}}.$$

This result can be obtained by applying **Stirling's formula**, which states that, for large n, we have $n! \approx \sqrt{2\pi n} n^n e^{-n}$. For $n = 12$ this formula already approximates the factorial with a relative error of less than one-thousandth, and the error decreases with each subsequent n.

The Stirling formula, which we used to approximate $n!$ for large n, is an example of what is called an **asymptotic formula**. In general, an asymptotic formula can be used to replace the value of a function with another function which is an approximation of the original, but is for some reason easier to calculate or manage. Asymptotic formulae are characterized by the fact that, the larger the value for which we are approximating the function, the better approximation we get. In particular, at infinity there is no error at all—that is why the Gaussian function that we obtained previously is an exact result, even though at some point we used Stirling's formula to approximate the factorial. In other words, $F_{as}(n)$ is an asymptotic approximation of $F(n)$ when

$$\lim_{n \to \infty} \frac{F(n) - F_{as}(n)}{F(n)} = 0.$$

The faster the left-hand side approaches zero, the better the formula. The relative error of Stirling's formula $((F(n) - F_{as}(n))/F(n))$ decreases proportionally to $1/n$.

The approximated formula clearly shows that the probability of returning to the starting-point decreases with the number of steps, and is equal to zero

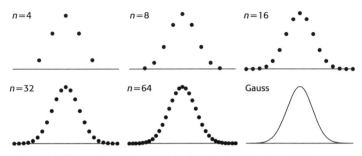

Fig. 4.3 The probabilities of reaching each slot in n-level Galton boards for $n = 4, 8, 16, 32,$ and 64. The last picture is the Gaussian distribution (when $n \to \infty$) given by the formula $\sqrt{1/\pi}\,\exp(-x^2)$.

at the limit when n reaches infinity. Applied to the coin-tossing interpretation, this means that the probability of getting an equal number of heads and tails diminishes with the number of tosses, which might seem paradoxical at first. It ceases being so, however, when we realize that the probability of obtaining *any other* ratio of heads to tails decreases as well, since more possibilities appear with each toss. In the random walk model, this means that the number of possible ending-points increases with the length of the journey, and so each particular point is less probable. Even though the probabilities of reaching each particular point approach zero, returning to the origin is still more probable than reaching any other specific point. In order to illustrate the probability distribution of reaching the point x at the limit when n approaches infinity, we may multiply the sequence of probabilities by a factor of $\sqrt{n/2}$. In order to keep the sum of probabilities equal to 1, we must shrink the x scale as well by dividing it by the same factor. What we get is the **Gauss curve** depicted in Fig. 4.3 and given by the formula $\sqrt{1/\pi}\,\exp(-x^2)$.

 Carl Friedrich Gauss (1777–1855); German mathematician, astronomer, and physicist.

The Gaussian distribution, also called the **normal distribution**, appears very often in nature. For example, if some property of a living organism is measured by a single number, then more often than not its distribution throughout the population will closely follow a scaled version of the Gauss curve. A popular example of this phenomenon is human height. If one takes a large enough group of people and plots the number of individuals of each height, then one will obtain a good approximation of the Gauss curve. Most people will be of some average height and only single individuals will fall into the extreme very short and very tall cases. Of course, the heights should be represented on a histogram; a perfectly exact measurement would not show much, since no two people are of exactly the same height. The normal distribution also applies to weight, shoe size, or even IQ, and it models the distributions of most properties of living organisms.

Similarly to the probability of reaching the starting-point, we can calculate the average distance traveled after n steps. This value is the sum of the probabilities of reaching each slot multiplied by the distance between that slot and the center. It turns out that the average traveled distance is proportional to $\sqrt{n}$. This holds not only in one, but in any number of dimensions. It can be observed in the process of diffusion. A drop of ink in a still glass of water will grow until the entire glass is colored. The radius of the drop grows proportionally to the square root of time.

Calculating the average distance traveled during a random walk is fairly complicated. It is, however, much easier to calculate the average squared distance. The square root of this value is the **standard deviation** σ of the particles' positions. The standard deviation of a set of values is defined as the square root of its **variance**. The variance, in turn, is the sum of the squared differences between the values and the average. In the case of the random walk, the average position is naturally the origin. The variance is, therefore, the average of the squared distances traveled. The standard deviation is a good approximation of the average distance from the center, though it is actually equal to it only when all of the distances are equal to each other.

Our previous analysis of the Galton board and the Pascal triangle shows that the probability of reaching the point s after n steps of a one-dimensional random walk is $(1/2^n) \binom{n}{(s+n)/2}$, provided that s is no less than $-n$ and no more than n, and that $s + n$ is an even integer (see Fig. 4.2). To simplify the notation we can denote $k = (s+n)/2$. Then, the squared distance from the center is the square of the position $s^2 = (2k - n)^2$, the probability of reaching this distance is $(1/2^n)\binom{n}{k}$, and the possible values of k are $0, 1, \ldots, n$. This leads to the following formula for the variance:

$$\sigma^2 = \sum_{k=0}^{n} (2k - n)^2 \frac{1}{2^n} \binom{n}{k}.$$

Some advanced sum manipulations can be used to show that the above sum evaluates to n, so the standard deviation in one dimension is equal to $\sigma = \sqrt{n}$. In the d-dimensional model, the squared distance from the center is calculated according to the Pythagorean theorem:

$$d^2 = \sum_{i=1}^{d} d_i^2.$$

In the above formula, d_i is the distance along the ith axis, i.e. d_i is the ith coordinate. Note that each step of the random walk is performed along exactly one of the d axes and changes exactly one of the d coordinates of the particle. Therefore, a sequence of n steps can be split into d subsequences, each subsequence containing parallel steps—n/d steps on average. Each of these subsequences models a one-dimensional random walk; so, according to the previous calculations, the average squared distance it yields is n/d. A sum of d components, each on average n/d, gives again n.

The program **Smoluchowski** illustrates the Brownian motion of a group of particles modeled by the random walk in one, two, or three dimensions. It calculates the standard deviation of the particles' positions and compares it with the square root of the time elapsed. The standard deviation is the average over the squares of the distances of each particle from the center, and it is a good approximation of the average distance.

Marian Smoluchowski (1872–1917); Polish physicist who, along with Einstein, developed the theory of Brownian motion. If it were not for his premature death of typhoid, he would most probably have been awarded the Nobel prize.

Sources of randomness

Let us now return to the important question of generating random numbers with a computer. The programs presented as illustrations to this and the previous chapter seem to behave in accordance with what we were able to derive using probability theory. This leads us to think that the random number generator used by these programs generates 'truly' random numbers. Unfortunately, this is not the case.

Computers do not generate random numbers, but only so-called **pseudo-random** numbers.

Computers are by design deterministic, which means that, if a computer is in a given state and behaves in a certain way, then it will behave in exactly the same way every time it is in that state. This property deprives computers of the ability to behave randomly and hence produce random data. What they can do, however, is take a number and apply a collection of mathematical functions to it in such a way that the result does not yield any similarity to the original number, and hence appears 'random'. The initial number is what is called a *seed* of the pseudo-random number generator. The resulting pseudo-random number can in turn be used as the seed to generate the next pseudo-random number, and so on. Depending on the quality of the algorithm used to generate each subsequent number, the resulting sequence can be a very good simulation of true randomness. Yet it must be stressed that the same seed always yields the same random number. Therefore, if the iteration is repeated a sufficient number of times, then eventually one will obtain the seed used at the beginning and the generator will start generating exactly the same numbers over again. In other words, a sequence resulting from pseudo-random number generation is always periodic. What is called the *period* of a random number generator is the period of the resulting sequence. The length of this period is one of the main measures of generator adequacy, though it is not the only one. Naturally, the numbers generated must also be uniformly distributed within a given range and give no evidence of the correlations between them. Different statistical tests can be used to discriminate between better and worse generators.

Most programming languages have a built-in random number generator with a period of rarely more than 2^{24}. This is suitable for most everyday uses, like picking random questions from a set to implement a trivia game, since it is hard to imagine that such a game would contain more than ten million different questions, and even harder that a player would be willing to answer them. However, even such a simple program as our Bernoulli program exposes the limits of a simple generator. No matter how long a sequence one would generate using the 2^{24}-period generator built into Microsoft® Visual Basic, one would *never* get more than twenty-five ones in a row, even though, as we showed earlier, such patterns should statistically appear in sequences billions of characters long.

The pseudo-random number generator used in all of our programs is called the *Mersenne twister* and is far more elaborate than the simple ones. Its period is $2^{19\,937} - 1$, which means that, if one used it to generate a billion random numbers a second, then it would still take 10^{5985} years (for all practical purposes this is eternity) before the sequence started repeating itself. The generator is also very fast and comes out very well in the tests which measure generator adequacy. That is why the results obtained from our programs agree with our theoretical calculations and do not give any hint of the fact that the numbers are not actually random. Yet we must still bear in mind that even a very good algorithm cannot generate truly random numbers.

The Mersenne twister was developed by Makoto Matsumoto and Takuji Nishimura. More information can be found on the homepage of Makoto Matsumoto at Hiroshima University.

Random numbers play a very important role in cryptography. Secure messages are encoded using a key which is typically some large number. Only one who knows the value of this key can decipher the message. There are systems which continuously transfer large amounts of secure data to different recipients; examples are the secure internet infrastructure which enables online shopping by supplying credit card details, or mobile phone networks routing calls which should not be eavesdropped. Such systems need a source of numbers to use for keys. The only way to prevent a hacker from guessing a key and deciphering a message not intended for him is to make the numbers random. Otherwise, if the keys follow a pattern, then a hacker could recognize this pattern and use it to predict the keys generated by the system. This is where even very good software random number generators fall short. Since each number is used to seed the generator for the next number, it is enough to know one of the keys to obtain all of the others.

For example, if an online store used such a generator to encode communication with its customers, then a hacker could browse their page and learn one key from the generator's sequence, since it would be the key used to communicate with him. Several other values would easily lead him to the exact algorithm used for generation, and from that moment on he could harvest all of the credit card details of the other customers. This scenario shows that a typical software generator is not what is called **cryptographically secure**.

One way to overcome this weakness is to apply some additional irreversible scrambling function to the random number (a so-called hashing function) to generate the key. In that way, knowledge of some set of keys does not necessarily lead to guessing the algorithm and the location in the generator period. Although for some purposes this yields a satisfactory degree of security, for very sensitive data any determinism in generating keys is still a threat.

We have already said that computers are fully deterministic, which would mean that they are unable to generate a truly cryptographically-secure, non-deterministic, sequence of values. Fortunately, this is not completely true. While the vital components of a computer, such as the processor and memory chips, act in a fully deterministic way (assuming that they are not damaged), other hardware components do not. Moreover, users of computers act non-deterministically as well. Computers are able to harvest the non-predictability which they are able to perceive and use it to make up for their own deterministic nature. They can collect data, such as the intervals between key-strokes, fluctuations in the motion of a disk drive, or noise gathered from a sensitive microphone, and use it to generate random numbers.

An alternative to gathering noise from peripheral devices is a hardware random number generator card—a card designed specifically for supplying a computer with true randomness. Such cards contain circuits which produce chaotic and unpredictable electronic noise. This noise is measured and translated into random bits. A good card of this sort can even cost several thousand dollars, which teaches us to never underestimate the value of randomness. As we can see, true randomness has a rare value in itself, even more so because it is inachievable either by man or by computer.

As we noted in the previous chapter, the decimal representations of some irrational numbers, such as π, $\sqrt{2}$ and e, are characterized by a uniform distribution of digits. It turns out, in fact, that such sequences have all of the qualities of a random sequence. Naturally, they are not cryptographically secure, since they follow a well-known pattern, but can serve as a very good, non-periodic, random source for other applications, such as the Monte Carlo algorithm described in the previous chapter.

Numbers possessing such a property are called **normal numbers**. Their formal definition is as follows. A number is normal in a given base if and only if every possible sequence appears in the infinite representation of that number in that base with equal probability. To illustrate the meaning of this definition, let us switch from the decimal system to the one with base 26. In the base 26 system we may use letters of the alphabet instead of digits to represent numbers. For example, we can define A to be the digit 0, B to be the digit 1, and so on; finally, Z would be the digit 25. For example, the decimal number 1371 would appear as CAT in this representation, since

$$(CAT)_{26} = (C)_{26} \cdot 26^2 + (A)_{26} \cdot 26^1 + (T)_{26} \cdot 26^0$$
$$= 2 \cdot 676 + 0 \cdot 26 + 19 = (1371)_{10}.$$

The infinite representation of π in the described system contains all possible texts, since π is normal at base 26. In particular, it contains the entire text of the *Encyclopedia Britannica* (disregarding spaces and punctuation) an infinite number of times. However, the probability of actually finding the text of the *Encyclopedia Britannica* is so low that one would need to search something like $10^{45\,000\,000}$ digits of π before finding its first occurrence. This quantity is so enormous that it cannot even be compared with anything.

5

Twenty questions

Probability and information

Information is one of the key concepts of the modern age. Storing and transporting information has become perhaps an even more important issue to mankind than storing and transporting physical goods. The organized and efficient flow of information between people all over the world, which modern technology makes possible, enables society to act as a whole in the pursuit of knowledge. It can help save lives and prevent disasters. It can also grant us access to an abundance of entertainment. All of this is just the tip of the iceberg of what processing and transmitting information means to us. In this chapter we will use the popular game of *Twenty questions* to illustrate the mathematical properties of information and, above all, we will learn how to measure information precisely.

One must stress that the mathematical theory of information does not deal with the meaning or significance which the information may carry. It deals only with the capacity of communication channels needed for transmitting a portion of information in a given amount of time. In line with that theory, the sentence, 'Humpty Dumpty sat on a wall, Humpty Dumpty had a great fall' can be directly compared with one of profound importance like, 'Energy curves space–time', since we take into account only the amount of information carried by these sentences and not the weight of their actual meaning.

Bits of information

We shall start by pointing out that information and uncertainty are two sides of the same coin. By gaining information we reduce uncertainty, and vice versa—by losing information we increase uncertainty. The greater the uncertainty regarding a result, the more information we gain on discovering it. Gaining information can be compared to filling a container with water. Water (information) fills the empty space (uncertainty) of the container. The volume of the container can be measured by the amount of water which it is capable of holding. Likewise, we can measure information and uncertainty in the same way—using the same units.

The unit of information is a **bit**. The term bit, originally introduced by John Wilder Tukey, actually comes from **binary digit**, but its association with something small is no coincidence. One bit is how much information we receive in answer to an **elementary question**, that is a question to which the answer can only be 'yes' or 'no'. Hence the bit is a very small unit; in fact, it denotes the smallest possible positive amount of information that can exist in nature.

 John Wilder Tukey (1915–2001); a mathematician at the University of Princeton; the first to use the word 'software' in print (1960) and one of the creators of the fast Fourier transform (FFT).

Since the bit is so small, very often we use a larger unit of information—the **byte**, equal to eight bits. Bytes in turn come in kilobytes (KB), megabytes (MB), gigabytes (GB), and terabytes (TB). Note that $1\,\text{KB} = 2^{10}$ bytes $= 1024$ bytes, and not 1000 bytes as is the case of grams in a kilogram, watts in a kilowatt, etc. This is underscored by the capital letter K in KB, and has been so defined since the base of 2 is much more natural in information theory than base 10. The same holds for the larger units, and hence $1\,\text{MB} = 2^{20}$ bytes $= 1\,048\,576$ bytes, $1\,\text{GB} = 2^{30}$ bytes $= 1\,073\,741\,824$ bytes, and $1\,\text{TB} = 2^{40}$ bytes $= 1\,099\,511\,627\,776$ bytes.

Many different media may be used to convey information; examples are sound, picture, motion, smell, and many others. However, the one we will concentrate on here is text or other information in digital form, since in this form the information content is easiest to measure.

A text is an ordered sequence of symbols (letters or characters). The set of all characters is called the alphabet. When writing down a natural number we most often use the alphabet composed of 0, 1, 2, 3, 4, 5, 6, 7, 8, and 9.

When writing words in a natural language we use the alphabet specific to that language.

The smaller the alphabet, the more we have to write to express the same thing. One can always translate a text written using a larger alphabet to a smaller alphabet, at the expense of extending the length of the text, and vice versa. We must, however, have at least two symbols in our alphabet, in order to supply a choice. If all we could do is to repeat the same symbol, then we could not transfer information.

The notation which uses an alphabet of two characters is called **binary**, and it is the one which we will most often consider from now on. The characters of the binary alphabet are customarily denoted by 0 and 1. Binary notation is native to computers. Music on a CD, a film on DVD, a mobile phone conversation, and all other digital data have this form. The main advantage of this notation for our purposes is that one binary character is one answer to an elementary question: 0 means 'no' and 1 means 'yes'. Therefore, each character carries exactly one bit of information.

There does exist a rarely used concept of unary notation, which is based on a single-letter alphabet. The unary representation of a natural number is this single letter repeated as many times as the number denotes. This does not apply to our speculations about information, since, in order to draw information from unary notation, we must know where it begins and ends, and hence introduce at least one extra symbol to the alphabet.

We will use the term *binary word* to denote a text in binary notation. As we said before, each character in this word is one bit of information and so the amount of information contained in the entire word is *proportional* to its length. This is of no surprise—a book which is twice as long holds (potentially!) twice as much information. In physical terms one would say that information is an extensive quantity, meaning that it is proportional to the size of its container, like the mass of a uniform body is proportional to its volume. Therefore, we arrive at our first observation:

$$\text{Information(Word)} = \text{Length(Word)},$$

where Word is a binary word, Length(Word) is the number of its characters, and Information(Word) is the information contained in it. We have set the proportionality constant in this formula equal to one, which means that we measure the information in bits—a binary word containing one character carries one unit of information.

It is easy to calculate that there are 2^N different binary words of length N:

$$2^{\text{Length(Word)}} = \text{DifferentWords}.$$

Taking the logarithm to base 2 of both sides of the above equation gives

$$\text{Length(Word)} = \log_2(\text{DifferentWords}).$$

Assuming that all words are equally probable, the occurrence probability p of a word is

$$p = \frac{1}{\text{DifferentWords}}.$$

From the above equations and properties of the logarithm ($\log(1/x) = -\log(x)$), we obtain the following formula for the information H_1 contained in a *single* word with occurrence probability p:

$$H_1 = -\log_2(p).$$

Claude Shannon (1916–2001); American mathematician and engineer; an employee of the renowned Bell Telephone Laboratories.

What we have just deduced is very similar to the measure of information introduced by Claude Shannon, who formulated the mathematical theory of communication. The year 1948, in which Shannon published his paper 'The mathematical theory of communication', is considered to be the birthdate of modern information theory. Nowadays, information theory has become a branch of computer science, but it still centers around one key formula–the **Shannon formula**, which expresses the information content of a message of length n in terms of the occurrence probabilities p_i of the symbols making up that message. The formula is

$$H = -\sum_{i=1}^{n} p_i \log_2(p_i).$$

Here H denotes the amount of information in bits. Since probabilities are always between zero and one, their logarithm is never positive (see Figs 5.1 and 5.2) and so the information H is always positive. We can, however, omit

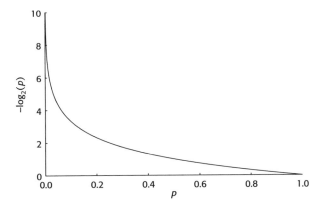

Fig. 5.1 The graph of $-\log_2(p)$.

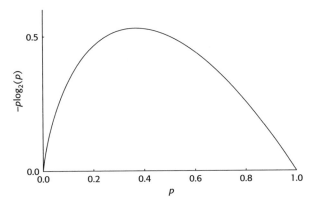

Fig. 5.2 The graph of $-p\log_2(p)$.

the minus sign in Shannon's formula and interpret H, which would then be a negative quantity, as the lack of information—uncertainty.

Due to the similarity of Shannon's formula to a formula for entropy in thermodynamics, the quantity denoted by H is often called the 'information entropy'.

So far we have used the binary alphabet to explain the meaning of the Shannon formula. How would this formula change if we use a longer alphabet? First, let us notice that, by increasing the number of letters from N to N^2, we shorten the length of each word by a factor of 2. For example, adding two letters, 2 and 3, to the binary alphabet enables us to write all of the four binary pairs, 00, 01, 10, and 11, as the single characters 0, 1, 2, and 3, respectively. Therefore, if the length of a word written in the binary alphabet is $2n$ or $2n-1$, then its new representation will have only n characters. As an illustration, we list below the number HM (one hundred million) written in alphabets that have $N = 2, 4, 8, 16,$ and 32 letters.

$N = 2$, HM $= 101111101011110000100000000$, length $= 27$,
$N = 4$, HM $= 11331132010000$, length $= 14$,
$N = 8$, HM $= 575360400$, length $= 9$,
$N = 10$, HM $= 100000000$, length $= 9$,
$N = 16$, HM $= 5f5e100$, length $= 7$,
$N = 32$, HM $= 2vbo80$, length $= 6$.

We clearly see the decrease in length by a factor of 2 when going from $N = 4$ to 16. Similarly, by increasing the number of characters from N to N^3, we reduce the length of a word by a factor of 3, since each triple of characters is now replaced by a single character. In general, when going from N to N^k, we decrease the length by a factor of k (or slightly less). Or, to put it differently, by increasing the size of the alphabet from N to M, we reduce the length of the words by a factor of $\log_N(M)$. This is confirmed by our example above. We could now repeat all of the arguments that led us to the Shannon formula, but this time taking an alphabet with N characters. The resulting formula for the information content (let us denote it by H_N) would be to the same form as before, except that the logarithm would be to the base N instead of the base 2. However, with the help of the chain rule for logarithms, we obtain $\log_N(p) = \log_N(2)\log_2(p)$. Thus, the new information measure H_N differs by a constant factor $\log_N(2) = 1/\log_2(N)$ from the measure based on the binary alphabet. Since H_N is measured in bigger units, we have to reduce H by a factor of $\log_2(N)$ to obtain the information content in these units:

$$H_N = -\frac{\sum_{i=1}^{n} p_i \log_2(p_i)}{\log_2(N)}.$$

Is it a dog?

The game of *Twenty questions* is a perfect illustration of the facts regarding information and probability sketched above. In this popular pastime, one of the players thinks of an object and the other player must guess what his opponent has in mind by asking only elementary questions, and therefore receiving answers of either 'yes' or 'no'. If he manages to do this using twenty or less questions, then he wins. An example of a match of twenty questions could be as follows.

—Is it a living being?
—Yes.

—Is it a person?

—No.

—Is it a mammal?

—Yes.

—Is it alive now?

—No.

—Is it fictional?

—Yes.

—Is it a dog?

—Yes.

—Does it come from a book?

—Yes.

—Is the book a children's book?

—Yes.

—Was the book written later than 1960?

—No.

—Is it Lassie?

—Yes.

We see in this example how one of the players was able to start from complete ignorance about the sought object and then successfully identify it, by asking only ten elementary questions and thus receiving exactly ten bits of information.

We will show that the measure of information contained within a word—the Shannon formula—is simply equal to the number of questions one needs to ask in order to guess this word. Needless to say, we mean elementary questions. For the time being we will assume that all words are equally probable. Since the word is written in binary notation, we can simply ask, 'Is the first symbol 1?', 'Is the second symbol 1?', etc., and then transform the resulting sequence of answers into the 0s and 1s of the word. A word of N symbols carries N bits of information, and following this pattern we receive N answers and hence N bits of information; so everything seems to be in order. However, for the benefit of future considerations, we shall describe a different approach to the word guessing problem.

Let us imagine that we have in front of us 2^N boxes and we know that exactly one of these boxes contains treasure. We have to ask questions that will lead us to finding out which box contains the treasure. Finding the treasure is equivalent to guessing the binary word, since there are 2^N different words and 2^N different boxes.

Initially, our knowledge of the treasure's position is none. It could be any one of the 2^N boxes. We could, of course, ask questions like, 'Is it in the

Initially we know nothing:

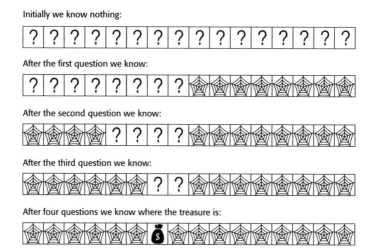

After the first question we know:

After the second question we know:

After the third question we know:

After four questions we know where the treasure is:

Fig. 5.3 The simplest strategy of asking questions.

first box?', 'Is it in the second box?', etc. However, most probably we would keep getting 'no' as the answer, and on average we would have to ask half as many questions as there are boxes before we find the treasure. This gives 2^{N-1} questions and is much more than the N questions which we know to be sufficient, so this strategy would be wasteful.

The optimal strategy in the case of all boxes being equally probable is to make sure that, with each question, we reduce the number of candidate boxes by a factor of two. That is to say, the first question we ask should be: is the treasure in the first half of the boxes? No matter what the answer, we only have $2^N/2 = 2^{N-1}$ boxes left to search, since after the first question we know which half the treasure is in. Figure 5.3 illustrates this strategy with an example where $N = 4$. In this case the questions and answers are as follows.

—Is the treasure in the first half of the boxes?
—Yes.
—Is the treasure in the first quarter of the boxes?
—No.
—Is the treasure in the third eighth of the boxes?
—No.
—Is the treasure in the seventh box?
—Yes.

Since each time we reduce our examined set by half, we can be sure that after exactly N questions we will arrive at the solution.

Reality is, however, usually more complex than illustrated by the above example and the halving strategy is not quite enough. This is because most often the probabilities of different configurations *are not the same*. Let us take as an example natural language. If when playing *Hangman* we are faced with an incomplete word of the form FA_E, then we do not know if it is FACE, FADE, FAKE, FAME, FANE, FARE, FATE, or FAZE. However, if we see QU_T then we can immediately be sure that the word is QUIT. This is because the pattern FA_E characterizes eight words and QU_T just one. In the first case we are uncertain regarding the word in question. Finding out the missing letter would eliminate this uncertainty and provide three bits of information, since it would identify one of the eight possible words. In the second case the third letter provides no information at all, since we already know the word from the remaining three letters. The more possibilities there are, the greater the uncertainty and the more information we gain by eliminating this uncertainty.

Let us go back to the treasure hunt example. Suppose that this time the treasure is in one of four boxes, but the probabilities of finding it in each of the boxes are *different*. Suppose that the probability that it is in the first box is 1/2, in the second 1/4, and 1/8 for both the third and fourth boxes. We thus have the following distribution of probabilities:

$$\text{distribution I}: p_1 = \frac{1}{2}, p_2 = \frac{1}{4}, p_3 = \frac{1}{8}, p_4 = \frac{1}{8}.$$

We could still use our halving strategy and ask exactly two questions, but this tactic is no longer optimal, since it does not utilize the information about the different probabilities. A better strategy would be the following.

—Is the treasure in the first box?

If the answer is yes then we are finished; otherwise, the next question should be as follows.

—Is the treasure in the second box?

If the answer is yes then we have found the treasure; otherwise, we ask one more question to decide between the third and fourth boxes.

This strategy might seem worse since in some cases it calls for asking three questions, as opposed to the two we are guaranteed by the halving strategy. However, if the treasure is in the first box then we find it after just one question, and it being in the first box is the most probable case. So on **average** we gain more than we lose.

We can calculate the average number of questions exactly in the most straightforward way—by considering all possible cases. For each case we take its probability p_i and multiply it by the value w_i associated with that case.

The sum of the results from all of the cases gives the average value W:

$$W = p_1 \cdot w_1 + p_2 \cdot w_2 + p_3 \cdot w_3 + \dots .$$

For example, the average number obtained when throwing a dice is

$$\frac{1}{6} \cdot 1 + \frac{1}{6} \cdot 2 + \frac{1}{6} \cdot 3 + \frac{1}{6} \cdot 4 + \frac{1}{6} \cdot 5 + \frac{1}{6} \cdot 6 = \frac{21}{6} = 3\frac{1}{2}. \qquad (5.1)$$

If we use this method to calculate the average number of questions needed in the previous example then we get

$$\text{Questions On Average} = \frac{1}{2} \cdot 1 + \frac{1}{4} \cdot 2 + \frac{1}{8} \cdot 3 + \frac{1}{8} \cdot 3 = \frac{7}{4} < 2.$$

So, in the case of distribution I, the second strategy proves to be better by an average of 1/4 question than the previous one. We arrived at this strategy intuitively by choosing the questions so as to keep the probabilities of receiving both possible answers equal. This is in fact the optimal strategy. In many cases, however, we are not able to find questions which ensure that the probabilities of both 'yes' and 'no' are exactly 1/2. An example could be the following distribution of probabilities for a set of six boxes:

$$\text{distribution II}: \quad p_1 = \frac{1}{3}, p_2 = \frac{1}{5}, p_3 = \frac{1}{5}, p_4 = \frac{2}{15}, p_5 = \frac{1}{15}, p_6 = \frac{1}{15}.$$

In this case we are unable to divide the probabilities into two sets of equal sums. We can, however, try to make the sums as close as possible to 1/2. Reasoning in this way brings us to the strategy presented in the form of a decision tree in Fig. 5.4. This strategy yields an average number of questions equal to 37/15. It is, however, *not* an optimal strategy for this distribution. An optimal strategy is depicted in Fig. 5.5—it gives an average of 36/15 questions; we will show later that this is the best we can get.

Building a strategy

There exists a method for producing an optimal decision tree when given a distribution of probabilities, and it is the following. We build the tree starting from the bottom—from the leaves. The leaves of a tree are the nodes with no children; in our case these are the points where we arrive at a conclusion and need not ask more questions. The leaves of a decision tree will always correspond to all of the possible conclusions—in this case to each of the boxes.

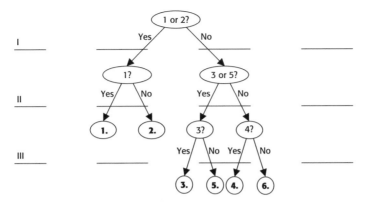

Fig. 5.4 The first strategy for distribution II.

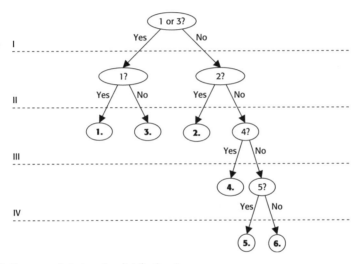

Fig. 5.5 The second strategy for distribution II.

For each of the leaves we know the exact probability that we will reach this leaf, since it is equal to the probability that the treasure is in the corresponding box. When building the tree we must keep track of the probabilities of reaching each node.

After we have created the leaves, the next step is to find two of the least probable nodes and 'glue' them together; we create a new node which corresponds to a question that combines the two least probable results. We must assign a probability to the new node as well. Reaching this node means that we have narrowed our search down to the two nodes in question, which in turn means that the treasure is in one of the two boxes corresponding to the leaves being glued. So the probability of reaching the new node is equal to the sum of the probabilities of reaching the two leaves. Once the nodes are glued, we can treat them as one. In the decision-tree language this would mean that we forget about the two glued leaves and take into account the new node instead. Even though the new node is not a leaf, it can be treated in the same way as the others, since it has an assigned probability. One just has to keep in mind that it corresponds not to one, but to more boxes.

In this fashion we keep gluing nodes (both the leaves and the resulting new nodes) until we are left with just one—the root of the tree, the first question to ask. Needless to say, the probability that we reach the root should come out as 1, since we always start from the root and therefore always reach it; and indeed it does, since it is the sum of all of the probabilities in the given distribution and these must always sum to 1 by definition.

To make it clearer, let us try to apply this algorithm to distribution II. We encourage readers to take a piece of paper and construct the tree with us. We have six boxes, and so six leaves. The distribution tells us how to assign probabilities to the leaves. Once this is done we pick the two least probable nodes. These are the nodes corresponding to boxes 5 and 6, both having a probability of $1/15$. So we glue them together and we are left with the leaves corresponding to 1, 2, 3, and 4, and a new node corresponding to 5 and 6 together, whose probability is $1/15+1/15 = 2/15$. Now the lowest probabilities are $2/15$ of leaf number 4 and $2/15$ of the new node. So we glue leaf 4 to the new node, thus creating a node corresponding to 4, 5, and 6 with a probability of $2/15+2/15 = 4/15$. Next we glue leaves 2 and 3 together to produce a node of probability $2/5$. We then glue leaf number 1 to the node corresponding to 4, 5, and 6, and finally we glue the two remaining nodes together. The resulting tree is depicted in Fig. 5.6. It is a different tree to the one in Fig. 5.5, but it still gives an average of $36/15$ questions. Often many optimal trees exist, but the above algorithm will always yield one of them.

The above algorithm is implemented by the **Huffman** program, which displays an optimal decision tree when given a set of probabilities.

David A. Huffman (1925–1999); professor at MIT and University of California Santa Cruz. As a graduate student at MIT, he invented the most popular lossless data-compression algorithm.

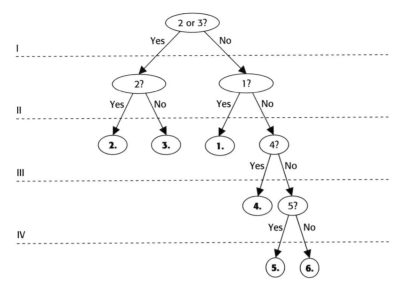

Fig. 5.6 The decision tree resulting from applying the Huffman algorithm to distribution II.

Presenting a strict mathematical proof that this method indeed always yields an optimal strategy is outside the scope of this book. We hope, however, to provide the readers with an intuitive feeling of why it works. We start from the leaves because these are the nodes which we know for sure will be in the tree. Then we have to start adding new nodes to create our tree. However, adding a node means that to reach this node we will have had to ask a question. The total number of asked questions is the quantity we want to minimize; so the answer is simple: we need to minimize the probability that we will reach the node where the question has to be asked. Obviously, the probability that we reach the node is the sum of the probabilities of reaching its children, since having passed through the node we go one way or the other. So we need to find two nodes with lowest total probability, and these will simply be the two least probable ones. We know that during the whole process we must add one less internal node than there are leaves, since each time we reduce the number of top-level nodes by one and our aim is to end up with a single top-level node—the root. So minimizing the number of internal nodes—questions asked—is not an issue; we have no control over it. All we can do is to minimize the probability that we reach such a node and actually have to ask the question.

The above algorithm was invented by David Huffman, and finding the optimal strategy for asking questions is just one of its applications. Its main use is in the field of data compression. In Chapter 9 we explain how it can be employed to compress text files.

The entropy bound

Let us now return to Shannon's entropy formula. What this formula actually tells us in terms of the treasure hunt is how much information (in bits) we get by finding out the position of the treasure, given a distribution of probabilities. This is also equal to the information we *lack* as we start the game, our initial uncertainty about the position of the treasure.

Let us try the simplest example. If all of the probabilities are 0 except for one which is equal to 1, then we know right away where the treasure is. We lack no information from the beginning and we gain no information by finding out that the treasure is indeed where it has no other choice than to be. So Shannon's formula should give zero, and one can indeed check that it does.

What might look questionable if we allow for zero probabilities is taking the logarithm of zero, a forbidden operation. The result of this logarithm will, however, be multiplied by zero anyway, so it will give zero in the end. (This is seen in Fig. 5.2, but readers with a sufficient knowledge of calculus can check that $\lim_{x \to 0} x \log(x) = 0$.) The logarithm of one is always zero in any base.

We have already said that the answer of 'yes' or 'no' gives us one bit of information. So the average number of questions that we need to ask to find the treasure is equal to the average information in bits, which we get by finding out the position of the treasure, and that information we lack at the beginning. So let us check if this agrees with our strategies. For an even distribution among 2^N boxes, we said that we need N questions. Shannon's formula gives us a sum of 2^N identical terms, each equal to $1/2^N \cdot \log_2(1/2^N) = 1/2^N \cdot (-N) = -N/2^N$. If we sum all of these 2^N terms and apply the minus then we get $-2^N \cdot (-N/2^N) = N$, as expected.

For distribution I we have

$$-\left(\frac{1}{2}\log_2\left(\frac{1}{2}\right) + \frac{1}{4}\log_2\left(\frac{1}{4}\right) + \frac{1}{8}\log_2\left(\frac{1}{8}\right) + \frac{1}{8}\log_2\left(\frac{1}{8}\right)\right)$$

$$= -\left(\frac{1}{2}\cdot(-1) + \frac{1}{4}\cdot(-2) + \frac{1}{8}\cdot(-3) + \frac{1}{8}\cdot(-3)\right)$$

$$= \frac{1}{2} + \frac{2}{4} + \frac{3}{8} + \frac{3}{8} = \frac{7}{4},$$

which is what we obtained by keeping the probabilities of 'yes' and 'no' equal.

The program **Shannon** calculates the information entropy according to Shannon's formula, given a distribution of probabilities.

If we apply Shannon's formula to distribution II, then the logarithms no longer come out rational, and so they are harder to calculate by hand. We can use the **Shannon** program to find out that the result is 2.365596. This is a bit less than the result we obtained, which was $36/15 = 2.4$. However, by looking at the way the average number of questions is calculated, we can see that if the probabilities are all multiples of $1/15$ then the average number of questions must also be a multiple of $1/15$. So the next best result we could get is $35/15 = 2.333333 < 2.365596$; however, this is a result that is impossible to achieve for the fundamental mathematical law of noiseless coding, which applied to our case is stated as follows.

No strategy yields an average number of questions lower than the value of H given by the Shannon formula measuring the amount of information gained by finding the answer.

It turns out that Shannon's universal formula applies more broadly than solely in the fields of information theory and communications. Apparently, human reaction time to events is proportional to the information contained in these events as measured by Shannon's formula. In 1953, an American psychologist Ray Hyman performed a series of experiments, during which he measured the subjects' reaction time to stimuli containing a variable amount of information. The title of Hyman's article 'Stimulus information as a determinant of reaction time' gives a very good summary of his discoveries. The subjects of the experiments were exposed to signals coming from eight lamps which lit up on a console. The lamps were labeled (bun, boo, bee, bor, bive, bix, beven, and bate) and the assignment was to recognize the signals and react by naming the lighted lamp. The preliminary stage of the experiment consisted of training the subject to connect each lamp with its name. Hyman found that if the signal was only of one type—like always the same lamp lighting up—then the subjects' reaction time depended only on the speed of transmission within the nervous system. However, with an increase in the variation of the signals, the reaction time increased by the time needed for the brain to process the information. Hyman showed that, if the signals consisted of the lighting up of N lamps occurring with probabilities

$p_1, p_2, p_3, \ldots, p_N$, then the average reaction time to such signals was proportional to the information the signals carried as measured by the Shannon formula.

 Our program **Hyman** performs a simplified version of Hyman's experiment and plots the user's reaction time as a function of the number of possibilities.

At the end of our discussion about information contained in texts, we would like to turn the reader's attention to an apparent paradox. Let us consider the infinite representations of normal numbers mentioned in Chapters 3 and 4. Such a representation contains all possible texts. One would therefore be inclined to assume that a sequence like that, for example the base-26 representation of π, contains an infinite amount of information. On the other hand, π is just a number which can be defined with simple formulae; these are formulae which can be easily memorized and which obviously hold a finite (though quite useful) amount of information. The catch is that the seemingly infinite amount of information encoded in the sequence is completely unreachable. It is true that the sequence contains all possible texts in its infinity, so it also contains the *Encyclopedia Britannica*, for example. But how would one find the *Encyclopedia Britannica* in the sequence if one did not know where it is or know exactly its content? If we knew its content then finding it in the sequence would not yield additional information. If we knew the position where the text of the *Encyclopedia Britannica* starts in the decimal representation of π, then the number representing the position would be so big that storing it would require much more space than the text of the *Encyclopedia Britannica* itself. In such a case the pointer to that position would be the bearer of all of the *Encyclopedia Britannica*'s information and not the number π in itself. The representation of π could in this case be regarded only as a tool in deciphering the data contained in the position number—a key to the encryption perhaps, but not a source of information in itself.

The simplest known, and historically the first, formula for calculating π was found in the seventeenth century and is $\pi = 4(1 - 1/3 + 1/5 - 1/7 + 1/9 - 1/11 + \ldots)$. Another notable formula was found by Srinivasa Ramanujan, a self-taught Indian mathematician, who astonished many university mathematicians with the originality and depth of his results. The formula is

$$\frac{1}{\pi} = \frac{\sqrt{8}}{9801} \sum_{n=0}^{\infty} \frac{(4n)!(1103 + 26309n)}{(n!)^4 396^{4n}}.$$

It might seem less elegant but it has a giant advantage over simpler formulae: adding each subsequent term to the sum increases the accuracy a hundred million times. To achieve the accuracy of just one term of Ramanujan's formula one would have to sum up a million terms of the seventeenth-century formula. Yet another remarkable formula for π was discovered in 1995 by David Bailey, Peter Borwein, and Simon Plouffe in the form of the following infinite series:

$$\pi = \sum_{n=0}^{\infty} \frac{1}{16^n} \left(\frac{4}{8n+1} - \frac{2}{8n+4} - \frac{1}{8n+5} - \frac{1}{8n+6} \right).$$

The unusual property of this formula is that with its help we can calculate the kth digit in the expansion of π, and we do not have to start the calculation from the very beginning. We do not need all prior digits, as is required in all previously known methods, to calculate any particular digit. There is a catch, though. The digits are not in decimal notation, but in hexadecimal notation.

The amount of *useful* information in π should therefore be defined as the shortest definition of this number. The same holds for all numbers and all other objects. In information theory the **complexity** of a problem is often defined by the length of the shortest algorithm solving that problem, and the complexity of a number is defined by the length of the shortest algorithm generating that number. This measure of complexity is named after one of its inventors, the Russian mathematician Andrei Nikolaevich Kolmogorov, and is now considered to be the standard approach for measuring what we have called the useful information contained in an object.

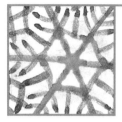

6

Snowflakes

The evolution of dynamical systems

Cellular automata, such as *The game of life* described in Chapter 2, belong to a larger family of mathematical 'models of reality'—the family of dynamical systems. A **dynamical system** is a mathematical object and a set of rules, which define how this object changes in time (evolution rules). The object can be anything, as long as it can be described by a set of numbers (hence the term *mathematical*), since only then can one compose the mathematical equations to describe the evolution rules for it. Sets of numbers, points, vectors, geometric figures, and such satisfy this criterion.

A cellular automaton is a dynamical system in which the object is a set of cells. The cells have some numerical properties associated with them (for example, 1 for an occupied cell and 0 for an empty cell) and the evolution rules state how these properties change from one generation to another. We shall call the set of values representing the mathematical object at a given point in time the **state** of the system.

A simple system

Before we go on to more complex dynamical systems, let us consider a very simple cellular automaton. This automaton models, in a very primitive fashion, the formation of a snowflake, which is a special case of crystal growth. A substance crystallizes primarily on contact with some other solid object, or an already-formed crystal. Therefore, a crystal needs something to begin forming around. In the case of snowflakes, it is typically a particle of dust or a bacterium. On contact with it, water turns into ice.

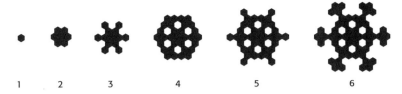

Fig. 6.1 The consecutive stages of development for a dynamical system resembling the formation of a snowflake.

Wilson Alwyn Bentley (1865–1931); American farmer who earned the nickname Snowflake for his passion for photographing flakes of snow. Starting in 1885, he took over 6000 such photographs throughout his life. It turned out that no two were identical. The collection of photographs was published in the form of an album.

Snowflakes have a hexagonal symmetry, which results from the shape of water molecules. Therefore, we will embed our model snowflake within a hexagonal lattice, and substitute all of the complex rules governing crystal growth with one very simple one: a cell 'crystallizes' when it has exactly one crystallized neighbor. The first six stages of such a system are depicted in Fig. 6.1.

The program **Conway** may be used to implement not only our classical *The game of life*, but also other cellular automata. It supports three types of two-dimensional grids (square, triangular, and hexagonal) and any set of rules in which the state of a cell depends only on its previous state and the number of living neighbors. It allows for the two possible cell states of alive and dead.

You can use **Conway** to model the growth of a snowflake, by choosing 'Edit rules' and picking the predefined set called 'Wolfram's snowflake' described in detail in Wolfram's book (Wolfram, 2002). The application then switches to the hexagonal lattice and turns on cell birth only in the case of exactly one neighbor. In this model living cells always survive to the next generation. After these settings are applied, all you need to do is place the initial 'speck of dust' somewhere and run the simulation.

The mathematical object in the case of this dynamical system is the set of crystallized cells. Each cell can be defined by two integers, namely its row number and its column number (see Fig. 6.2). So, from the mathematical point of view, the first three states of our snowflake could be as shown in Fig. 6.3.

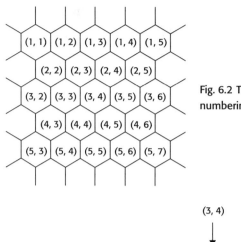

Fig. 6.2 The row and column coordinates numbering cells on a hexagonal lattice.

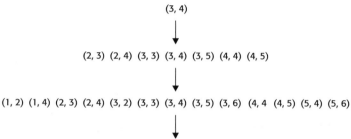

Fig. 6.3 The first three stages of snowflake growth, represented as pairs of numbers corresponding to the rows and columns of crystallized cells.

A purely mathematical description of the rules for the snowflake could be written as follows:

$$n1(r,c) = (r-1,c), \quad n2(r,c) = (r,c+1), \quad n3(r,c) = (r+1,c+1),$$

$$n4(r,c) = (r+1,c), \quad n5(r,c) = (r,c-1), \quad n6(r,c) = (r-1,c-1),$$

$$S_{t+1} = S_t \cup \{(r,c)|\exists n_i(r,c) \in St\}.$$

In the above formulae, (r,c) is an ordered pair of coordinates numbering the rows and columns, $n_i(r,c)$ is the pair of coordinates of the ith neighbor (counted clockwise), and S_t is the set of crystallized cells at time t. Here $\exists$ stands for 'there exists exactly one'.

Iteration

The development in time of a dynamical system can be described by an **evolution equation**. The evolution equation is an **iteration formula**

$$X_{t+1} = F(X_t),$$

where X_t represents the state of the system at time t. **Iterating** (which comes from the Latin term *iteratio*—repetition) a function means applying it over and over again. We take an object and apply the function to it; then we apply the function to the result, and so on. For this to be possible, the outcome of the function must be of the same type as its argument. This is not always the case. An example of a function that cannot be iterated could be the grade average function, one that takes a student as input and returns the average of his grades. The function's argument is of type student and its outcome is of type number. Since these two types are completely different, we cannot apply the function again to the outcome, and so we cannot iterate it.

The states X_t can be any mathematical objects, as long as they are of the same type. They can be single numbers or more complex structures defined by many numbers. In the case of *The game of life* the state X_t is the set of living cells, and in the case of the snowflake the state X_e is the set of crystallized cells. In both cases X_t can be represented as a set of number pairs.

The assumption that the next state of the system, X_{t+1}, depends only on the previous state, X_t, is not a severe restriction. We may also easily accommodate into this scheme those cases in which the next state depends on a *finite* number of previous states. This is done by simply including in the definition of each state all of the relevant previous states.

The next chapter illustrates how a very simple iteration can yield surprisingly complex results. First, however, let us look at iteration formulae from the general point of view. Each such formula can be used to construct an infinite sequence of terms X_t, provided that the first element X_0 is specified. In other words, the process depicted in Fig. 6.4 can be repeated an infinite number of times. Often the resulting sequence depends on the choice of X_0, but sometimes X_t converges to the same value regardless of the first element. Sumerian mathematicians were aware of this 4000 years ago and they used it to calculate square roots.

The method for calculating square roots is one of the oldest algorithms. It is based on the observation that if $r = \sqrt{a}$ then $r = a/r$. If $a/r > r$ then r is too small to be the square root of a; if $a/r < r$ then it is too large. Therefore,

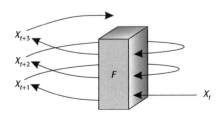

Fig. 6.4 The iterating machine F—a 'black box' transforming X_t into X_{t+1}.

Table 6.1 The calculation of $\sqrt{2}$ using the iteration formula and $x_0 = 2$.

$x_0 = 2.000000000000000000000000000000000000000000000000$
$x_1 = \underline{1.}500000000000000000000000000000000000000000000000$
$x_2 = \underline{1.41}66666666666666666666666666666666666666666666667$
$x_3 = \underline{1.41421}568627450980392156862745098039215686274509804$
$x_4 = \underline{1.414213562}3746899106262955788901349101165596221574$
$x_5 = \underline{1.41421356237309504880}1689623502530243614981 92577620$
$x_6 = \underline{1.41421356237309504880168872420969807856967187537}723$
$\sqrt{2} = 1.41421356237309504880168872420969807856967187537695$

setting r to be the average of a/r and r always brings the value of r closer to the square root of a. This leads to the following iteration formula:

$$x_{t+1} = \frac{1}{2}\left(x_t + \frac{a}{x_t}\right). \qquad (6.1)$$

It is easy to check that, for every positive x_0, the sequence built from this iteration converges to $\sqrt{a}$. For negative x_0 it converges to $-\sqrt{a}$. The proof is based on calculating the following limit as $t \to \infty$ of both sides of the iteration formula:

$$x_\infty = \frac{1}{2}\left(x_\infty + \frac{a}{x_\infty}\right). \qquad (6.2)$$

The above equation can be easily reduced to $x_\infty^2 = a$, and so its solutions are $x_\infty = \sqrt{a}$ and $x_\infty = -\sqrt{a}$.

This extremely simple dynamical system is a great tool for calculating square roots. Setting $a = 2$ and $x_0 = 2$ and applying the iteration six times gives $\sqrt{2}$ to an accuracy of forty-eight digits after the decimal point (see Table 6.1). The seventh iteration corrects the approximation by another forty-eight digits. Starting with 3, 4, or 5 yields very similar results. With the increase of x_0, however, more iterations are needed to achieve the same accuracy.

Attraction

The former, very simple square-root algorithm illustrates two important concepts of dynamical systems, namely fixed points and attractors. A **fixed point** of the function F is an element x such that $F(x) = x$. Notice that in the context of iterations it is an element which produces a constant sequence. The fixed points of the square-root iteration were $\sqrt{a}$ and $\sqrt{a}$.

An **attractor**, in turn, is an element to which the sequence resulting from iteration converges. In the case of the algorithm for square roots, the fixed points were also the attractors of the iteration, though this is not always the case. For example, 1 and $-1/2$ are fixed points of the iteration $x_{t+1} = 2x_t^2 - 1$, but neither of them is an attractor.

Another important concept is the **basin of attraction**. A basin of attraction is the set of elements which, when iterated, converge to an attractor. In the previous example all of the positive numbers composed the basin of the attractor $\sqrt{a}$ and the negative numbers were the basin of $-\sqrt{a}$.

A similar, but more general method for finding roots was invented by Sir Isaac Newton. It finds the roots of any given function. A root of a function is the value for which the function gives zero. In particular, the roots of the function $f(x) = x^2 - a$ are $\sqrt{a}$ and $-\sqrt{a}$. Newton's method works with complex numbers as well as real ones and, similarly to the previous algorithm, gives different roots for different starting-points. Using Newton's method to find the complex third roots of 1 and plotting the basins of attraction of the three different roots gives an infinitely complex fractal structure. Fractals will be described in detail in Chapter 8.

Sir Isaac Newton (1643–1727); English physicist and mathematician; considered by many to be the greatest physicist of all time.

You can use the program **Mandelbrot** to explore the basins of attraction of the complex third roots of 1, among other fractals. The application assigns each point on the complex plane (a plane on which each point corresponds to one complex number) one of three colors, depending on which basin of attraction the point belongs to.

Basing his derivations on the Taylor series, Newton came up with the iteration formula

$$z_{t+1} = z_t - \frac{f(z_t)}{f'(z_t)}$$

for finding a root of the function $f_{(z)}$. Here $f'(z)$ is the derivative of $f_{(z)}$ and z may be complex. In particular, finding the complex third roots of 1 is equivalent to finding the zeros of the function $f(z) = z^3 - 1$, which leads to the following iteration formula for generating Newton's fractal:

$$z_{t+1} = \frac{2}{3}z_t + \frac{1}{3z_t^2}.$$

If z is split into its real and imaginary parts x and y, respectively, by assigning $z = x + iy$, then it can be interpreted as a point on a plane. Iterating each point using the above formula, observing which of the three roots it converges to, and assigning one of three colors to it accordingly, results in a plot of Newton's fractal. The formula, split into real and imaginary components, is as follows:

$$x_{t+1} = \frac{2}{3}x_t + \frac{1}{3}\frac{x_t^2 - y_t^2}{(x_t^2 + y_t^2)^2},$$

$$y_{t+1} = \frac{2}{3}y_t - \frac{1}{3}\frac{2x_t y_t}{(x_t^2 + y_t^2)^2}.$$

Of course, Newton did not know anything about fractals and did not even realize how interesting the basins of attraction of his iteration method could be. Newton's fractal was described three centuries later along with other similar mathematical objects such as the famous Mandelbrot set.

Dogs and fleas

The rules exemplified so far unambiguously determine the next state of the system X_{t+1} given the current state X_t. In other words, each of the dynamical systems presented so far represents a **deterministic** process.

If the rules contain elements of chance, then we are dealing with a **stochastic** (non-deterministic) process. The simplest stochastic process (and a completely stochastic one) is throwing a coin. Each stage is completely random and does not depend at all on the previous one.

There exist systems which contain both deterministic and stochastic elements. A good example is the 'dogs and fleas' problem, which was formulated by Paul Ehrenfest to explain the seeming irreversibility of physical processes. The problem models the flea infestation of each of two dogs; let us call them Ace and Butch. At first, only Ace is infested with n fleas and Butch is flea-free. Once the two dogs meet, fleas start jumping from one to the other. Let us suppose that in each step exactly one flea decides to change dog. Which flea jumps in each step is random, with the assumption that each flea is equally anxious to jump. Therefore, the probability Pa that a flea will decide to jump from Ace to Butch is proportional to the number of fleas on Ace (a), and the probability P_b of a flea jumping from Butch to Ace is proportional to the number of fleas on Butch ($n - a$). Consequently, $P_a = a/n$ and $P_b = 1 - a/n$. At first, $P_a = 1$ and $P_b = 0$, since all of the fleas are on Ace ($a = n$) and no fleas can come from Butch.

Paul Ehrenfest (1880–1933); theoretical physicist born in Vienna; professor at Leiden University (Holland).

The state of the dogs-and-fleas dynamical system can be described using the single number a, since the total number of fleas (n) is constant (we can call n a *parameter* of the system). If we let a_t denote the number of fleas on Ace at time t, then the evolution equation of the system is as follows:

$$a_{t+1} = \begin{cases} a_{t-1} & \text{with probability } a_t/n, \\ a_{t+1} & \text{with probability } 1 - a_t/n. \end{cases}$$

The deterministic part of the system is the distribution of fleas among the dogs; the stochastic part is the choice of flea to move in each step. As we can see, each subsequent state of the system depends both on the deterministic value and the random event.

You can use the program **Ehrenfest** to simulate the evolution of the dogs-and-fleas system for different values of n.

The complete evolution of the system can be split into two phases. At first, the majority of fleas reside on Ace, so most fleas jump from Ace to Butch and the number of fleas on Ace dramatically decreases. This decrease is exponential and can be modeled by the function $a_t = (n/2)e^{-2t/n}$. Once the distribution of fleas between the two dogs evens out, it stays more or less equal to $n/2$, notwithstanding some natural fluctuations.

The irreversibility which Ehrenfest aimed to show is the fact that, after the system reaches equilibrium, it will never return to its original state, or at least the probability that it will is extremely small—this probability becomes smaller as the total number of fleas increases. This phenomenon pertains to many physical processes. Let us imagine a container with two chambers—one empty and the other filled with gas. When the wall separating the chambers is removed, the particles of gas quickly fill the entire container. In theory, nothing stands in the way of the gas completely moving back to its original chamber after some time, but the probability of this happening is unimaginably small. The program **Ehrenfest** shows that even for $n = 20$ the gathering of all fleas on one dog is extremely unlikely; for gas in a container (the number of particles being of the order of 10^{26}) it is for all practical purposes impossible.

The dogs-and-fleas system contains both deterministic and probabilistic elements. In many such cases it is useful to describe the evolution of a system using probability alone. Then, it is the probability itself that is subject to evolution in time. Equations describing the evolution of a probability distribution play an important role in many fields ranging from physics to social sciences. In fact, their role is so significant that they have been named **master equations**.

The master equation for the dogs-and-fleas system is extremely simple and it can be obtained from the formulae given above. Owing to the probabilistic nature of the flea jumps, only the average number of fleas can be predicted. The number of fleas on Ace decreases or increases by 1 from the value a_t with probabilities a_t/n and $1 - a_t/n$, respectively. Therefore, the average number of fleas $a_t + 1$ on Ace at time $t + 1$ is equal to

$$a_{t+1} = (a_t - 1)\frac{a_t}{n} + (a_t + 1)\left(1 - \frac{a_t}{n}\right) = a_t + 1 - 2\frac{a_t}{n}.$$

Dividing this equation by n and identifying a_t/n with the probability $P_a(t)$ to find a flea on Ace, we obtain the following evolution equation for this probability:

$$P_a(t + 1) = P_a(t) + \frac{1 - 2P_a(t)}{n}.$$

This is a very simple master equation. Its fixed point and also the attractor are equal to 1/2. It may be easily checked, even with a calculator, that the basin of attraction for this iteration is the whole allowed range of P_a, from 0 to 1 (without the end-points).

7

The Lorenz butterfly

Deterministic chaos

Webster's new encyclopedic dictionary defines **chaos** as a 'state of utter confusion'. This is the everyday sense of the word; however, since the 1970s *chaos* has gained a whole new meaning in the scientific domain, due to the rapid advance in computer modeling. In the scientific context, *chaos* is accompanied by the adjective *deterministic*, meaning 'predictable' or 'definable'. 'Deterministic chaos' seems to be an oxymoron—a paradox compressed into two words—since the adjective *deterministic* contradicts the traditional meaning of chaos. Deterministic chaos is a fully repeatable phenomenon. A computer simulation of deterministic chaos will always give the same results. There is, however, an element of this process which justifies the use of the word chaos. It is the extremely rapid increase in computational power needed to reconstruct a chaotic process.

Logistic iterations

To illustrate the phenomenon of deterministic chaos we will consider **logistic iteration**. The logistic iteration was introduced a long time ago by Pierre-François Verhulst, but it was Mitchell Feigenbaum's discovery which made it the paradigm of chaos-generating models.

Pierre-François Verhulst (1804–1849); Belgian sociologist and mathematician.

Mitchell Feigenbaum (1945–); American theoretical physicist. Using the first programmable calculator (HP-65) produced by Hewlett-Packard, he discovered universal laws governing deterministic chaos.

Logistic iterations have been used by biologists for many years as a simple model of the size of a population over time. A typical population conforms to the general rules of growth with limited resources. It grows due to reproduction and shrinks due to death. The number of newborn organisms in each unit of time is proportional to the overall size of the population (p)—the more parents, the more children. If we denote the reproduction coefficient by α, then this dependence is as follows:

$$p_{t+1} = p_t + \alpha p_t = (1 + \alpha)p_t.$$

The number of deaths depends on the size of the population as well (the more organisms, the more potential victims), but it also depends on the number of other organisms in the environment. Organisms consume resources, such as food, and in this way cause the death of other organisms. In this simple model the other organisms are simply the rest of the population, so the number of deaths is proportional to the size of the population in two different ways. In other words, it is proportional to the *square* of the population size. Let β denote the death rate. Introducing the death factor into our formula produces

$$p_{t+1} = (1 + \alpha)p_t - \beta p_t^2.$$

Since the units in which p is measured have not yet been specified, we can adjust them in order to eliminate one of the constants. We do this by substituting $x(1 + \alpha)/\beta$ for p and denoting $a = 1 + \alpha$. Finally, we get the following simple logistic iteration formula:

$$x_{t+1} = ax_t(1 - x_t).$$

The above formula is just as simple as the square-root generator described in the previous chapter, yet it reveals an unexpected complexity. Depending on the parameter a, we observe very diverse behaviors. For example, for $a = 1.1$ and $x_0 = 0.5$, the sequence steadily approaches a fixed point of 0.0909091 and reaches it after 132 steps (see Fig. 7.1(a)). In fact, an almost identical behavior can be observed for any value of $x_0 \in (0, 1)$ (see Fig. 7.2). Figure 7.1 shows,

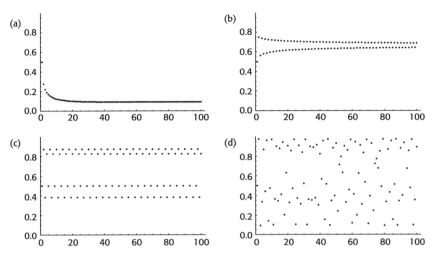

Fig. 7.1 The first hundred iterations of the logistic formula $x_{t+1} = ax_t(1 - x_t)$ for $x_0 = 0.5$ and (a) $a = 1.1$, (b) $a = 3$, (c) $a = 3.5$, and (d) $a = 3.9$.

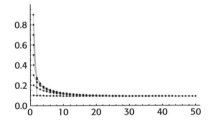

Fig. 7.2 The first fifty iterations of the logistic formula $x_{t+1} = 1.1x_t(1 - x_t)$ for different values of x_0.

however, that the iteration does not always converge to a fixed point. When $a = 3$ it oscillates between two values. The values slowly come closer and closer together, but they never become one. Setting $a = 3.5$ leads almost instantly to a stable oscillation between the four values 0.38282, 0.500884, 0.826941, and 0.874997.

Setting $a = 3.9$ yields a result that is indeed worthy of the term 'chaos'. At first glance the sequence seems to be a completely random scattering of points. A closer look (see Fig. 7.3) shows that it is not completely random. All of the values lie between 0.0950625 and 0.975, and their density is evidently higher at the edges as well as around the value 3.5. Other than that, however, the scattering is completely chaotic. Moreover, unlike the convergent and oscillating cases, this sequence is highly sensitive to the choice of x_0. A difference of 1/1000 in x_0 can yield a difference of 0.8 (practically the entire range) in x_{100}. A similar degree of sensitivity can be observed with respect to the parameter a.

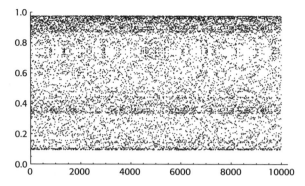

Fig. 7.3 The first 10 000 iterations of the logistic formula $x_{t+1} = 3.9\,x_t(1-x_t)$.

Butterfly's wings

It is the extreme sensitivity to initial data that characterizes chaos in the mathematical sense of the word. A model in which the tiniest disturbance in the input leads to completely different results renders computers helpless. Every computer is a finite machine and can only store numbers with a finite accuracy, so almost every number it manipulates is rounded. The rounding error is typically of the order of 10^{-16}, which is acceptable in most cases. However, in the examples shown previously any error, no matter how tiny, alters the results dramatically. In the case of smaller errors, more iterations have to be performed until the difference becomes apparent, but it always appears sooner or later.

This sensitivity of the logistic iteration for some values of a is, in fact, so high that the results depend on the particular computer used to perform the calculation, on the software applied, and on the way the iteration is programmed. To illustrate this, we tried to calculate x_{100} using the same computer (with a Pentium® 4 processor) and using the same software (Mathematica®), but we entered the iteration formula in the following two ways:

$$x_{t+1} = 3.9\,x_t(1-x_t),$$
$$x_{t+1} = 3.9\,x_t - 3.9x_t^2.$$

From the mathematical point of view, both formulae are exactly the same and so the results of both calculations should be the same as well. However, the two produced sequences already exhibit a difference of the order of 10^{-17} after the first two iterations (for $x_0 = 0.5$). The difference grows exponentially, and after seventy-five steps the two sequences hold practically no resemblance to each other, as seen in Fig. 7.4.

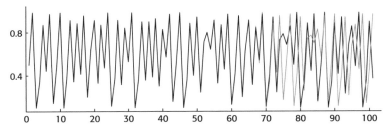

Fig. 7.4 Plots of the logistic iterations $x_{t+1} = 3.9\,x_t(1 - x_t)$ and $x_{t+1} = 3.9\,x_t - 3.9\,x_t^2$. Even though the functions are the same, the difference between the rounding error in both cases accumulates and becomes apparent after approximately seventy iterations.

Figure 7.4 clearly shows that any results for x_{100} computed numerically for this type of iteration cannot be trusted. The accumulated error is so great that the obtained values hold no resemblance to the correct result. Of course, we may program a computer to compute exact values of each number. For example, for $a = 39/10$ and for $x_0 = 1/2$ (as in our previous example) we obtain the sequence

$$\frac{1}{2}, \frac{39}{40}, \frac{1521}{16000}, \frac{858879801}{2560000000}, \frac{56981253341773815561}{6553600000000000000000},$$
$$\frac{190109472776760869780369661041446615955881}{4294967296000000000000000000000000000000000000}, \ldots$$

The results obtained here are exact, but the space needed for each subsequent value grows exponentially since the number of digits in the numerator and in the denominator roughly doubles with each step of the iteration. Thus, after one hundred iterations we would need something like 2^{100} bytes of storage space. All of the computers on Earth could store only a negligible fraction of that number.

Is there no way, therefore, to examine the logistic iteration with a computer? Luckily, there is still something we can do. It is true that the numbers which a machine produces given this type of problem have little in common with the true result. However, many interesting traits can be observed when studying the *statistical* properties of these numbers. As Fig. 7.3 shows, even the very random behaving versions of the iteration follow a well-defined pattern. To further study this pattern, let us construct a map of points according to the following algorithm. Take a list of a values ranging from 2.8 to 4 and differing by 0.004. For each a, run the logistic iteration for 200 steps. Discard the first 150 values of x_t (for $t = 0, 1, 2, \ldots, 149$) and plot the remaining points (a, x_t). The result of this procedure is shown in Fig. 7.5. Even though the exact positions of the points in the chaotic regions of the map depend on the details

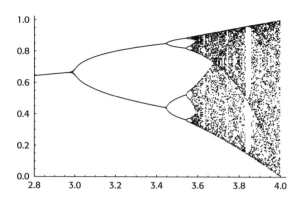

Fig. 7.5 A map of the behavior of the logistic iteration $x_{t+1} = ax_t(1 - x_t)$. The parameter a changes along the horizontal axis from 2.8 to 4. The points for each value of a correspond to $x_{151} \ldots x_{200}$.

of the procedure used to make the calculations, the overall appearance of the map is always the same.

Use the program **Feigenbaum** to explore the chaotic map described in the above paragraph. The program allows the exploration of different regions of the plot (for various values of a and x_t and with varying accuracy) and the control of the number of points plotted and discarded. We may also choose the iteration function.

We can use **Feigenbaum** to check that the statistical behavior of the logistic iteration is always the same, even though the individual points may differ from computer to computer. Unlike the specific results, the general aspect of the chaos map is not sensitive to changes in the input and even to modifications to the iteration function. Substituting $x_{t+1} = a \sin(x_t)$ for the logistic formula does nothing to change the characteristic, infinitely complex, branching aspect of the plot, even though the shape of the corresponding map is slightly different.

The phenomenon of tiny differences in x_0 causing huge differences in the values of x_t further down the iteration path is called the **butterfly effect**. The term comes from Edward Lorenz's article 'Can the flap of a butterfly's wing stir up a tornado in Texas?'. Lorenz stumbled upon chaos while using a simple meteorological model to predict the weather. In order to repeat only the second phase of a long computation, Lorenz entered a set of data printed by the computer midway during its first run. To his great surprise, the scientist observed a gigantic difference in the two sets of results which should have been identical. It turned out that the printed numbers were rounded versions of the numbers the computer kept in memory. This tiny difference in the input caused massive differences in the final result. Lorenz's chance discovery of this butterfly effect turned out to be one of the milestones in the development of chaos theory.

Edward Norton Lorenz (1917–); American meteorologist and mathematician; retired professor of Massachusetts Institute of Technology.

The program **Lorenz** models the meteorological phenomenon examined by Edward Lorenz. It plots a three-dimensional position as a function of time. The program illustrates the butterfly effect, since a tiny change in the initial position makes a large difference in the future ones.

The equations used by Lorenz (the ones yielding the evolution illustrated by the Lorenz program) and the logistic formula are strongly related. **Lorenz equations** are differential, but they can be approximated by the following set of three iteration formulae for the three coordinates:

$$x_{t+1} = x_t + 10(y_t - x_t)\epsilon,$$

$$y_{t+1} = y_t + (28x_t - y_t - x_t z_t)\epsilon,$$

$$z_{t+1} = z_t + \left(-\frac{8}{3}z_t + x_t y_t\right)\epsilon.$$

In the above formulae, ϵ is a small parameter defining the distance that the point travels in one step of the iteration. It is worth noting that chaotic behavior (at least in models with a finite number of dimensions) is caused by the occurrence of nonlinear terms in the iterations. In this case, these terms are the product of x and z values in the second equation and the product of x and y values in the third equation. The nonlinearity in the logistic formula is the square of x. Such terms make the formula for each subsequent x_t, as a function of x_0, an order of magnitude more complicated. The following equations illustrate this by expanding the dependencies given by the logistic formula for the first three iterations (note also the dependence on the parameter a):

$$x_1 = ax_0 - ax_0^2,$$

$$x_2 = a^2 x_0 - a^2 x_0^2 - a^3 x_0^2 + 2a^3 x_0^3 - a^3 x_0^4,$$

$$x_3 = a^3 x_0 - a^3 x_0^2 - a^4 x_0^2 - a^5 x_0^2 + 2a^4 x_0^3 + 2a^5 x_0^3 + 2a^6 x_0^3$$
$$- a^4 x_0^4 - a^5 x_0^4 - 6a^6 x_0^4 - a^7 x_0^4 + 6a^6 x_0^5 + 4a^7 x_0^5 - 2a^6 x_0^6$$
$$- 6a^7 x_0^6 + 4a^7 x_0^7 - a^7 x_0^8,$$

$$x_4 = a^4 x_0 + \cdots - a^9 x_0^{16}.$$

The formula for x_4 in terms of x_0 is a sixteenth-order polynomial with 63 terms. The formula for x_5 has 293 terms, with some coefficients exceeding 100 000. This complexity makes it impossible to store each value exactly, which causes the rounding errors to have a colossal impact on the results, thus leading to chaos. Not every nonlinearity leads to growing complexity and chaos. For example, the nonlinear **rational function** $f(x) = (ax + b)/(cx + d)$ maintains its form under iteration and hence does not exhibit any chaotic behavior.

To conclude the discussion about chaos, we would like to observe that the butterfly effect lies at the root of many events which we call **random**. The final result of throwing a dice depends on the position of the hand throwing it, on the air resistance, on the base that the die falls on, and on many other factors. The result appears random because we are not able to take into account all of these factors with sufficient accuracy. Even the tiniest bump on the table and the most imperceptible move of the wrist affect the position in which the die finally lands. It would be reasonable to assume that chaos lies at the root of all random phenomena.

8

From Cantor to Mandelbrot

Self-similarity and fractals

The notion of similarity is one of the most elementary, yet key, ideas of geometry. Two geometrical figures are **similar** if and only if we can obtain one from the other by means of shifting, scaling, and rotating. **Self-similarity** is a concept closely related to that of similarity, but it pertains to just one figure. Here the term figure means a set of points on a plane or in space.

> A figure is self-similar when it is composed solely of elements similar to it.

At first glance the above definition might seem inachievable, for how can a part be a reproduction of the whole? This doubt disappears in the face of the notion of infinity. Since geometrical figures can contain infinite numbers of points, nothing stands in the way of a one-to-one mapping between the points in such a figure and the points in just a part of it. Let us take as an example a line segment. Every segment is composed of two smaller segments appended to each other. The smaller segments are identical to the original segment except for the fact that they are scaled by a factor of 1/2 and one is shifted to the end of the other. Hence, a segment is self-similar. Likewise, filled triangles, parallelograms, and cubes, as well as many other filled geometric figures, are self-similar. These examples are, however, pretty trivial. Self-similar figures are interesting when their self-similarity brings about an infinitely complex fine-grained structure.

Cantor's set

The simplest, and one of the oldest, interesting self-similar structures is a one-dimensional set of points described in 1883 by Georg Cantor. The construction of this set begins with the closed segment [0, 1]. (Square brackets denote the fact that the set is closed—that it contains its end-points.) Next, we remove the open middle subset (1/3, 2/3), leaving two closed segments [0, 1/3] and [2/3, 1], each one-third of the original length. (Parentheses stand for an open set—one without end-points.) If we apply the middle-removing procedure to the obtained two segments, then we get four segments, each one-ninth of the length of the original. Upon repeating this process infinitely many times, we obtain the **Cantor set**. The first six stages of this construction are depicted in Fig. 8.1.

> The Cantor set never ceases to baffle mathematicians with its array of paradoxical properties. One of these is the fact that its total length is 0, yet it contains exactly as many points as the full [0, 1] segment! This seems unbelievable, yet note that a single point does not contribute anything to the length of a segment, since it is infinitely small and its length is zero. So, no matter how many points there are, their total length will be zero as long as the points are all disconnected from each other. To prove that the total length of the Cantor set is zero, it is enough to notice that with each step of the construction the length decreases by one-third and, as we must all remember from high school,
>
> $$\lim_{n \to \infty} \left(\frac{2}{3}\right)^n = 0.$$
>
> The tricky part is proving that there are as many numbers in the Cantor set as there are in the [0, 1] segment. This is done by demonstrating a one-to-one mapping between the two sets. In fact, demonstrating such a mapping is the only way one can show the equivalence in size of two infinite sets. It is intuitive that

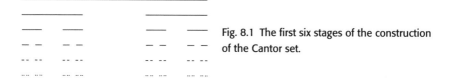

Fig. 8.1 The first six stages of the construction of the Cantor set.

Georg Cantor (1845–1918); German mathematician; father to a branch of mathematics known as set theory.

such a one-to-one correspondence between the elements of two sets implies their equality in size. What is less intuitive is the fact that we can remove elements from a set and still leave the set the same size—this is possible only with respect to infinite sets.

The key part of the proof is showing that the numbers forming the Cantor set are those which do not contain the digit 1 in their base-3 representations—the nth step of the construction removes numbers with 1 at the nth position following the decimal point. For example, removing the first middle segment, i.e. (.1, .2) in base-3 notation, removes all numbers with 1 at the first position after the dot. Actually, the number .1 is not removed, but .1 can be represented as .022222...—from the mathematical point of view, this is equivalent. Similarly, we can write .222222... instead of 1. Therefore, substituting all of the 2s in the base-3 representations of the Cantor set with 1s and interpreting the results as binary fractions gives all of the possible numbers from the [0, 1] segment. Therefore, both of the sets contain an equal number of elements—the number of infinite sequences of two different characters.

The Cantor set is self-similar. It is composed of two Cantor sets, each one-third of its length, with a void one-third of its size in the middle. The former sentence does not only demonstrate the self-similarity of the set. It may also serve was its definition, since the Cantor set is the largest one-dimensional structure which is composed of two copies of itself, each one-third in length, with void one-third of its size in the middle.

A definition which contains the term being defined is called a **recursive definition**. In some cases a recursive definition may prove to be the shortest, the most elegant, and the clearest of all. Recursive definitions may be used to define not only self-similar geometrical figures, but all sorts of other mathematical objects. For example, we may define the factorial function by stating that $n! = n \times (n - 1)!$ and $0! = 1$. This definition, unlike the one used for the Cantor set, contains an additional, not recursive component. If we interpret the definition as a recipe for calculating factorials, then this additional component tells us when to stop. The definition of the Cantor set contains just one component, since its construction continues indefinitely.

The program **Cantor** supplies a graphical interface for recursively defining two-dimensional self-similar figures. It allows a picture to be built out of transformed copies of itself. The transformations allowed in the program do not necessarily preserve the similarity of objects, since changing the aspect ratio and skewing is also allowed. This opens the way for a greater variety of interesting structures. The program comes with a set of sample designs, including the figures described in this chapter as well as a collection of some popular fractals.

Other self-similar structures

A similar construction to that resulting in a Cantor set may be used to construct the **Koch curve**. Again, we begin with a single segment and with each step we remove its middle part. This time, however, we replace the removed segment with two sides of an equilateral triangle, as shown in Fig. 8.2. Again, we must continue this replacing forever to obtain the fractal figure.

Helge von Koch (1870–1924); Swedish mathematician known mainly for the fractal curve named after him.

Applying the Koch curve construction to an equilateral triangle rather than to a single segment generates the **Koch snowflake** (or Koch island) depicted in Fig. 8.3. What is interesting about this shape is its ratio of area to circumference. With each step of the construction the perimeter of the figure increases by one-third, so the circumference of the finished snowflake is infinite. Yet we

Fig. 8.2 The first five stages of the construction of the Koch curve.

Fig. 8.3 The Koch snowflake.

Fig. 8.4 The Sierpinski triangle.

can clearly see that the shape can be inscribed into a finite circle, and so its area is limited.

Perhaps the most famous self-similar figure is the **Sierpinski triangle**, also known as the Sierpinski gasket or Sierpinski sieve. Its construction begins with a single filled triangle. As the first step, we divide this triangle into four parts by connecting the centers of its sides and we then remove the middle piece. This leaves three smaller triangles touching only by their corners. Applying this process indefinitely to the obtained triangles results in the Sierpinski triangle illustrated in Fig. 8.4.

Waclaw Sierpinski (1882–1969); Polish mathematician.

The Sierpinski triangle appears in an astounding number of contexts. For example, it is the result of marking the odd numbers in the Pascal triangle (described in Chapter 4), as shown in Fig. 8.5. Of course, the complete Sierpinski triangle comes from coloring the complete, infinite Pascal triangle. Each subsequent row of the Pascal triangle is formed by adding two values from the previous row, but we may modify this rule to take into account only the parity of the numbers, i.e. whether they are even or odd. Only the sum of one even and one odd number gives an odd number. This observation leads to the description of a simple one-dimensional cellular automaton with two cell states and the rule that a cell is marked if and only if it had two different neighbors in the previous step. Starting from a single marked cell, the subsequent states of this automaton form line after line of the Sierpinski triangle.

Sierpinski's triangle may also be obtained by means of a random walk variation, a phenomenon which illustrates how a stochastic process can

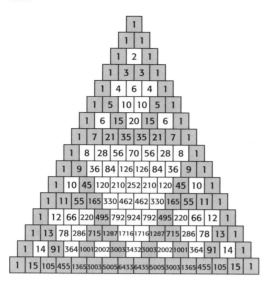

Fig. 8.5 Odd numbers in the Pascal triangle form the Sierpinski triangle.

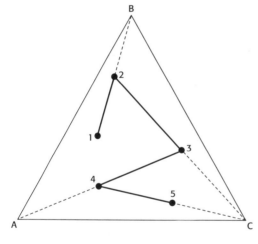

Fig. 8.6 A random walk within a triangle visits points belonging to the Sierpinski triangle.

produce a predictable result. Imagine a flea jumping around between three pegs. With each jump, the flea selects one of the pegs at random and jumps towards it, decreasing the distance between it and the peg by one-half (see Fig. 8.6). Wherever the flea lands it leaves a spot. The pattern of spots produced by the flea paints the Sierpinski triangle.

Use the program **Sierpinski** to plot the Sierpinski triangle by means of a random walk. With this program you may also modify the number and position of the pegs, as well as the fraction of the distance covered by the flea with each jump, to generate different fractal-like patterns.

Fig. 8.7 A fourth-level Menger sponge.

Variations of the Sierpinski triangle include Sierpinski's carpet, formed by removing the center part of a square, and the-Menger sponge, formed by removing the center and the centers of the walls of a cube (see Fig. 8.7). Countless other figures in any number of dimensions can be constructed in a similar fashion. All of them are self-similar and are usually called **fractals**, the term coined by Benoit Mandelbrot.

Other fractals

 Benoit B. Mandelbrot (1924–); mathematician born in Poland; studied in France; eventually settled in the USA.

Benoit Mandelbrot contributed greatly to the popularization of fractals, his most renowned discovery being the **Mandelbrot set**. Peitgen *et al.* (1992)

wrote of the Mandelbrot set as being not only the most popular fractal, but the most popular object of modern mathematics. The appeal of the structure undoubtedly lies in the spectacular beauty of the visualizations associated with it and in the simplicity of the rules which define it.

The Mandelbrot set exists on a two-dimensional plane. Each point of the plane may be used as the starting-element of a simple iteration to determine whether it belongs to the set or not. If the iteration causes the point to move further and further away from the origin, eventually escaping to infinity, then the starting-point does not belong to the set. If, however, the iteration always leaves the point in the vicinity of the origin, then the starting-point belongs to the set.

Points on a two-dimensional plane may be treated as single complex numbers, with one coordinate corresponding to their real part and the other to their imaginary part. In this interpretation the iteration used for defining the Mandelbrot set is

$$z_{t+1} = z_t^2 + z_0,$$

where z_t are the complex numbers corresponding to the sequence of points. Without the use of complex numbers and by labeling the iterating points as (x_t, y_t), the formulae are

$$x_{t+1} = x_t^2 - y_t^2 + x_0,$$
$$y_{t+1} = 2x_t y_t + y_0.$$

Julia sets are defined analogously to the Mandelbrot set, but with the iteration

$$z_{t+1} = z_t^2 + c,$$

where c is some constant. There is a different Julia set for each different value of c. Curiously, Julia sets resemble parts of the Mandelbrot set in the vicinity of the point c. The point c is sometimes referred to as the location of the Julia set.

Figure 8.8 contains a plot of the Mandelbrot set, a shape often referred to as the Mandelbrot bug. This picture does not do justice to the figure, as the beauty of the Mandelbrot set lies in its infinite complexity, a complexity which becomes apparent when exploring different regions of the structure at different magnifications. It is also worthwhile enriching the plots of the set with additional colors. The points belonging to the set are customarily colored black. The points outside, however, may be assigned different colors depending on the number of iterations it takes before the iterated point leaves the region defined by a circle of radius 2 centered at the origin. It has been

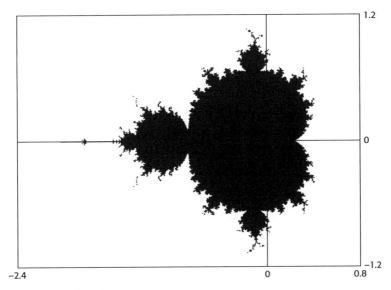

Fig. 8.8 The Mandelbrot bug.

proven that, once a point escapes from this circle, it will escape to infinity, and thus the starting-point will prove to lie outside the set.

The program **Mandelbrot** provides a tool for exploring different regions of the Mandelbrot set at different magnifications. It also allows the exploration of Julia sets and the linking of them to their location interpreted as a point of the Mandelbrot set. The application can also serve as a tool for examining the infinite complexity of Newton's fractal, described in Chapter 6.

Strictly speaking, the Mandelbrot set is not self-similar. It does contain an infinite variety of smaller 'bugs' inside it, yet each of these is slightly different from the original. The Mandelbrot set cannot be defined in terms of itself, as was the case of the fractals mentioned earlier, yet it is still a fractal in the sense that it is infinitely complex.

Fractional dimensions

The precise definition of a fractal is based on the concept of **fractal dimension**. There are several definitions of the notion of dimension. The one we are most used to is the **topological dimension**, defined as the number of coordinates needed to navigate within the object in question. For example, we need just one number to specify the location on a line—hence, a line is one-dimensional. Analogously, two numbers are needed to specify the location on a plane and three numbers in space, so a plane is two-dimensional and space is three-dimensional.

As follows from its definition, the topological dimension is always an integer. This has proven to be an over-simplification in the case of fractals, which exhibit mathematical properties not quite matching any integral number of dimensions. New definition of dimension have been formulated to describe these structures, definitions which allow dimensionality to be any non-negative real number.

The measure of dimensionality most commonly applied to fractals is the **box dimension**, also called the capacity dimension. The concept of box dimension comes from the observation that the dimension of an object is linked to the relationship between the number of tiny boxes needed to cover the object and the linear size of those boxes. The boxes are squares or cubes depending on the topological dimension of the space in which the fractal is embedded. The question is: how many such boxes does one need to completely cover the described object? This naturally depends on the size of the box and the scale of the object, but, as it turns out, also on the fractal dimension of the object. The relationship can be expressed as follows:

$$N = \frac{C}{s^d}.$$

In the above expression, s is the length of a side of each of the boxes, d is the dimension, N is the number of boxes needed, and C is a constant which corresponds to the scale of the object.

Let us take as an example a full unit segment. We need just one box of size 1 to cover it, but two boxes of size 1/2, three boxes of size 1/3, four boxes of size 1/4, and so on. In this case $N = 1/s^1$, so the box dimension of a segment is 1, which agrees with the topological dimension. The case of a unit square is different; we still need just one unit box to cover it, but four boxes of size 1/2, sixteen boxes of size 1/4, and so on. The number of squares needed grows proportionally to the inverse of the *square* of the box size, and hence the dimension is 2. Figure 8.9(a, b) illustrates these derivations. In a similar fashion, we may deduce that the box dimension of a cube is 3. In fact, the box dimension always matches the topological one in the case of smooth, non-fractal objects.

Figure 8.9(c) demonstrates how to calculate the box dimension of the Sierpinski triangle. In order to cover this fractal figure we need more boxes than for the segment, but less than for the square. Decreasing the box size by a factor of two triples the numbers of boxes needed to cover the figure. From this observation it follows that the box dimension of the Sierpinski triangle is

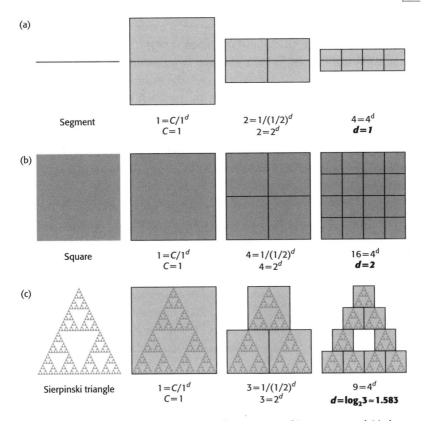

Fig. 8.9 Calculating the box dimension of (a) a line segment, (b) a square, and (c) the Sierpinski triangle.

$\log_2 3 \approx 1.585$. We leave as an exercise for the reader the proof that the box dimension of the Cantor set is $\log_3 2 \approx 0.6309$.

> The above examples were chosen so as to immediately yield the exact result. In general, however, the calculated value of d may vary depending on the box size, due to the fact that N is always an integer. The formal definition requires the calculation of the value of d in the limit when the box size reaches zero.

Fractals in real life

Fractal objects are not just picturesque illustrations of various mathematical concepts. They have recently proven to be quite adequate models of the world around us. In his book *The fractal geometry of nature*, published in 1982, Benoit Mandelbrot points out that natural phenomena such as the shape of a cloud or the path of lightning appear to be closer to a fractal than to a classical smooth figure. Of course, the 'fractals' which appear in nature are not mathematically perfect, but they have a great deal in common with the mathematical objects described in this chapter. A famous example is the length of the Great Britain coast. The closer the coastline is examined, the more twists and turns it has— just like the Koch curve; and, similarly to the Koch curve, the length of the coast turns out to be longer the more accurately we decide to measure it. Unlike in the case of the Koch curve, a coastline can only be measured with finite accuracy; yet, in both cases it is important to bear in mind that the result of such a measurement is correlated with the accuracy used and may differ from what we expect.

What has always fascinated scientists about fractal objects is how a very simple formula can yield visually very complex patterns. This gave rise to the idea that perhaps the process can be reversed—an image being translated into a few simple formulae. This process is called **fractal compression** and was researched as well as patented by Professor Michael Barnsley of the University of Wales, Swansea. Sadly, it turned out to be quite a problem to effciently find the self-similarity in an image, and fractal compression never saw the light of day as an image-compressing algorithm. Fractal recipes, however, prove very effective in such cases as the rendering of realistic terrain or trees in two as well as three dimensions. It is enough to look at Barnsley's famous fern leaf (a sample in the **Cantor** program) to see just how closely some fractals resemble reality.

9

Typing monkeys

Statistical linguistics

'How often might a Man, after he had jumbled a Set of Letters in a Bag, fling them out upon the Ground before they would fall into an exact Poem?' These words were used by the seventeenth-century archbishop John Tillotson to illustrate the fact that improbable occurrences are acts of God. A more modern quotation on the subject comes from the English physicist Arthur Eddington, who said during one of his popular addresses, 'If an army of monkeys were strumming on typewriters they *might* write all the books in the British Museum'. Both of these passages, which we have found in Bennett (1976), show that the difference between the written word and a random set of letters has been the subject of contemplation for a long time. In this chapter we will show how a computer may be used to make this comparison, to automatically recognize the language—or even the author—of a text, and what can be done to increase the chances of the typing monkeys producing a work of literature.

Statistical analysis of texts

Let us begin with the problem of automatic language recognition. We want to find an algorithm which when given a text determines the language it is written in. Naturally, a language which uses a character set different from all of the others is easy to tell apart. The question is how to differentiate between languages which use the same characters, or are written using these characters following some agreed upon transcription method. We may, of course, try to look up the words of the text in different dictionaries and see which dictionary yields the largest number of matches. This procedure, however, is slow and requires a lot of storage space to keep all of the dictionaries. The more extensive the list of languages under consideration, the more impractical this method gets.

Does a natural language have some other unique characteristic, other than a complete list of its words—a characteristic which is easy to remember and verify? It turns out that it does. This characteristic is the frequency distribution of the individual letters. Such information is taken into account, for example, when designing the game of *Scrabbler*®. In this game the letters which appear more often have fewer points associated with them. Different language versions of *Scrabble*® have different letter weights. Sometimes these weights differ dramatically. For example, the letter Z is one of the two most valuable letters (10 points) in the English *Scrabble*® and one of the least valuable (1 point) in the Polish version. This is evidence of the fact that a text containing a large percentage of Zs is much more likely to be found in Polish than in English.

The knowledge of the letter distribution in a text is enough to determine the language it is written in, provided that the text is suffciently long. An enhanced recognition algorithm may take into account not only the individual letter frequencies, but the frequencies of the different pairs of letters, triplets, etc. This increases the space needed to store the language descriptions, but also increases the probability of a correct guess in the case of a short text. For example, the English language has a very high percentage of TH pairs, while German is full of SCH triplets.

The program **Poe** analyzes a given text by calculating the occurrences of single letters, pairs, and triplets. The user may specify which characters are to be taken into account and which are equivalent. The single letter and pair occurrence data may be saved and used for later language recognition. The application comes with several language definitions based on famous literary works in different languages.

The computer-aided statistical analysis of texts may also be used to compare different texts written in the same language. In this case, the analysis is usually at the word level, rather than at the letter level. Counting the occurrences of different words in a text, as well as their different combinations, may be helpful in identifying the author of the text. Every writer's style is different and hence

Edgar Allan Poe (1809–1849); famous American poet, writer, and literary critic. The main character of his 1843 short story The golden bug deciphers an encrypted message on the basis of letter occurrence statistics. Ironically, Poe got the statistics wrong, placing the letter T as the tenth most common one, while, as our program shows, it is the second most common one (the letter E is the first).

different writers tend to use some words or combinations of words more often. On the basis of a statistical analysis, we may even detect an influence of one author on another.

Another use for automatic text analysis sprang up in recent years: spam detection. Spam is the popular word for e-mail advertisements which are sent without the consent (and even against the wish) of the receiver. Electronic mail is used on a very wide scale and is practically free. At the same time it is a relatively recent invention and the laws and regulations pertaining to other types of mail have yet to be sufficiently adjusted to fit this new area. All of this gives rise to an avalanche of junk e-mail messages flooding our mailboxes every day. Filtering out the important messages is a conceptually easy task for a human being, but with large amounts of spam it is a tedious and annoying one. This is where automatic spam-filtering programs can be very helpful. Such a program performs a statistical analysis of the words contained in an e-mail, and classifies the letter by comparing its content to data obtained earlier from an analysis of samples of regular and spam mail. For example, an e-mail containing 'NAKED CELEBRITIES ABSOLUTELY FREE!!!' will most likely get filtered out. However, spam authors aware of anti-spam programs often compose their message so as not to use words characteristic of spam. They might, for instance, replace the text mentioned earlier with 'NUDE STARS TOTALLY GRATIS'. That in turn leads to the update of the program's heuristic data and the 'vicious circle' can continue endlessly.

Calculating the occurrence probabilities of pairs, triplets, quadruplets, and so on of letters, words, or other elements in a sequence is a special case of examining the **correlations** between these elements. The word correlation ('co' + 'relation') means 'mutual relationship'. In this case the correlations tell us how the occurrences of different letters or words are related to the occurrences of other letters or words in the direct neighborhood. For example, in an English text the letter Q is almost always followed by a U. (Only devoted *Scrabble*® players are aware of the few exotic exceptions, such as 'qaid' or 'qoph'.) One would say that there is a strong correlation between the occurrence of a Q and that of a U, or that the occurrence of a Q is correlated with the occurrence of a U. The correlations between different letters are different in every language, while the correlations between different words may vary depending on the author, subject matter, the century the text originated in, and other factors.

Occurrence probabilities are useful when generating random text. In many fields of science, ranging from genetics to linguistics, we need to compare actual data with some sort of random text. A completely random text is often 'too random' in such cases—we want something arbitrary, yet something that still somewhat resembles the data that we are analyzing. We may generate a random text which tries to match the occurrence probabilities of the characters in the text we are analyzing. Or we may go further and try to match pair occurrence probabilities, those of triplets, quadruplets, etc. The longer the 'tuple' whose occurrence distribution we match, the more the random text will resemble the one analyzed.

The procedure for generating such a random text is simple. Suppose that we want to generate a sequence of characters for which triplet occurrence probabilities resemble, on average, those in a given text. First, for every pair occurring in the text, we calculate the probability of getting some letter z directly following the pair xy, for any letters x, y, and z. Then we generate a random pair, taking into account the pair occurrence probabilities in the original text. We generate the next character taking into account the probability distribution of characters occurring after the pair we have just generated. We use the last two letters of the obtained triplet to generate the next letter in the same fashion, and so on. This procedure can be generalized into the one where we take into account any n previous letters, where n is a natural number.

 The program **Poe** may be used to generate a random text, of any given length, whose triplet occurrence is modeled after a given text using the procedure described above.

It is surprising how much information is actually stored in the three-letter correlations of a natural text. The following random text generated from *Hamlet* is based on just these rules: 'the does th my artis fring come to this'; it is almost in English. Moreover, given a lengthier sample, one can easily tell if the random text was generated based on *Hamlet* or, say, on *Oliver Twist*.

Compression

Letter occurrence statistics are useful when compressing files. A computer file is typically a sequence of characters and each character occupies one byte of disk space. There are two types of compression, namely lossless and lossy. In the process of lossy compression, some information is purposely discarded in order to minimize storage space. A lossy compression algorithm must be aware of the sort of data that it is compressing in order to know which parts it may discard. For example, if the data being compressed is music, then it may discard sounds which are too low or too high to be audible by the human ear. We shall restrict our considerations, however, to lossless compression—the one which works with all types of data, is fully reversible, and keeps all of the information contained in the original file.

As we mentioned earlier, each character in a file occupies one byte (eight bits) of space. We cannot hope to store each of these characters in a smaller number of bits, for example seven, since a file can potentially contain all possible eight-bit combinations, i.e. all 256 characters. So where can we make the savings and how can the letter statistics help? The answer is simple. We may store more common characters using less bits and rarer ones using more bits. Since more common characters are, by definition, those which occur more often in the file, we will be saving space more often than not and the outcome should occupy less space than the original, without actually losing any information. This simple idea does not, however, immediately translate into a compression algorithm, since it leaves a few technical difficulties to be solved. These are as follows.

- Since different characters are coded using different numbers of bits, how do we know, when decompressing the file, when to stop reading a character? For example, after having read the bits 101, how can we know if it means character 101 or if it is the beginning of character 1011?
- Given the table of character occurrences in a file, how can we assign the variable bit codes to the different characters so as to make sure that the assignment is optimal?
- Since the character codes are different for each compressed file (since they depend on the character statistics) they must be stored along with the compressed file for later decompression. How do we store them efficiently?

The answer to the first problem is a **prefix code**. We must assign the bit codes to the characters in such a way that if 101 corresponds to a character, then no bit combination starting with 101 may correspond to a different character. In other words, no code may be the prefix of a different code. Notice that the same holds for telephone numbers. The telephone number 9114321 cannot be a valid number in the United States, since, before it could be dialed to the end, its three-digit prefix would cause a connection with the emergency services. A coding where no code is the prefix of another is called a prefix code (although a more logical name would be the prefix-free code).

The second problem has already been addressed in this book, though in a different form. Notice that the process of decompressing a file is a lot like finding out information by asking questions. Initially, we know nothing of what the file contains, so we read the first bit. A bit is like an answer to a yes/no question. The first bit decreases our uncertainty of what the file contains by providing some information on what the first character stored in the file may be. However, one bit is usually not enough to determine the character unambiguously, so we must read another bit from the file (ask another question), and so on. When we can finally be certain of what the first character is, we may write it down on the side and repeat this process for the second, third, and all remaining characters.

Readers who have read Chapter 5 of this book carefully might already see the analogy between finding an optimal question-asking strategy and finding an optimal prefix-code variable-bit coding. In fact, it is exactly the same problem. In the first case we are given the probability distribution of the possible answers. In the second case we are given the probability distribution (based on occurrences) of the different characters. In the first case we want to be able to arrive at the solution by asking questions and receiving in answer one bit of information at a time. In the second case we want to find out what character it is by reading from a file and receiving one bit of information at a time. In both cases we want to minimize the average number of questions asked (bits read). Therefore, calculating character occurrence probabilities based on their actual occurrences in a file and applying Huffman's algorithm (described in Chapter 5) to them yields an optimal coding. We must simply interpret the path from the root to a leaf as the bit code of the character corresponding to this leaf, where the 'yes' and 'no' branches are interpreted as 1 and 0 bits, respectively. In fact, Huffman's motivation for inventing this algorithm in the first place was data compression, not a passion for playing *Twenty questions*.

 The program **Huffman** can be used to compress and decompress files using Huffman's algorithm. The program allows the inspection of the code tree produced in the process and gives detailed data on what the compressed file contains.

The third problem, namely that of storing a tree, may be addressed in several ways. Probably the most optimal method is implemented in the **Huffman** program. We refer the reader to the 'Huffman implementation notes' section of the *Modeling reality* help file for details on how it works. For the benefit of the example below, we will just mention that a code tree encoded using this algorithm takes up exactly $10n - 1$ bits, where n is the number of different characters in the file being compressed.

Another minor issue is the fact that, on the PC platform, files cannot end in the middle of a byte. Therefore, when we have finished coding, we must fill the remaining space of the last byte with bits. This presents a problem when decoding, since one might interpret the extra bits as a new character. To overcome this problem the size of the original file is stored in the header of the compressed file. A header is useful anyway for recognizing a file compressed by this particular program.

The complete data compression algorithm, as implemented by our program, is best illustrated by an example. Let us try to compress the following fragment of a well-known nursery rhyme:

Humpty Dumpty sat on a wall Humpty Dumpty had a great fall

The text is fifty-eight characters in all (counting spaces, of course), but there are only twenty different types of characters in the text. Some characters (such as the letter g) occur only once in the text, while others (such as the space or the letter a) are much more frequent. We can apply the Huffman algorithm described in Chapter 5 to create an optimal coding tree based on these occurrences, or we can let the **Huffman** program do it for us. What results may vary depending on the order in which equally-frequent characters are evaluated and on how left and right are defined, but it will always be an optimal coding. The outcome as produced by the **Huffman** program is depicted in Fig. 9.1. The leaves contain different characters as well as the number of their occurrences in the text, and the branches are labeled with different bits. Of course, it does

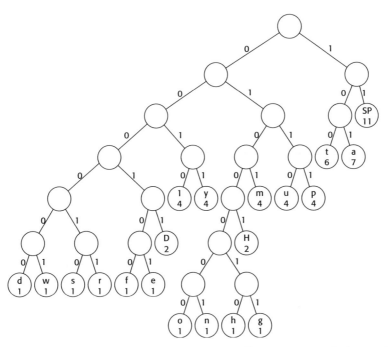

Fig. 9.1 The coding tree for 'Humpty Dumpty sat on a wall Humpty Dumpty had a great fall' produced by the Huffman algorithm.

not matter how we assign the bits to the branches as long as each internal node has two different branches leading down out of it.

To translate a character into a code, we simply follow the path to its corresponding leaf from the root, writing down the bits corresponding to the branches traversed on the way. For example, the code for H is 01001 and the code for u is 0110. To translate a code into a character, we start from the root and follow the branches corresponding to the bits in the code. When we reach a leaf, we have decoded a full character and may return to the root to start decoding a new one.

> The full text 'Humpty Dumpty sat on a wall Humpty Dumpty had a great fall' codes into the following sequence of bits:
>
> 01001011001010111100001111000110110010101111000011110000101011001101000001
> 00001111011100000110100100010110100101100101011110000111100011011001010101111
> 00001111010001010100000011101110100011000011000101101100110001001010010010010
>
> Readers may check that this bit sequence can be unambiguously decoded back into the text using the tree from Fig. 9.1.

The most frequent character, the space, has been assigned a code which is only two bits long. This decreases the length of the output file by six bits for each of the eleven occurrences of the space. In fact, each character is coded using less than eight bits, since the low number of different characters used in the very short sample allowed for it. Usually, however, one will find characters coded using more than eight bits. This does not bother us since these characters will be the ones which occur relatively rarely in the file.

By substituting the characters in the file with the bits corresponding to their codes, we are able to decrease its size from $8 \times 58 = 464$ bits to only 225 bits. As we mentioned earlier, however, the compressed file must be appended with the code tree, which is $10 \times 20 - 1 = 199$ bits, and a header, which in this case is 472 bits. As a result of this, the 'compressed' file is almost twice the size of the original, but this is only because the original file was so short that the header overhead alone accounted for more than one-half of the output. However, for large files the header/tree information remains almost unchanged, and therefore the savings can be great, especially for text files. The complete works of Shakespeare in text format consume 5.5 megabytes of disk space, while they require not much more than 3 megabytes when compressed using our program.

Commercial programs for archiving do an even better job and are able to compress Shakespeare to a mere 2 megabytes. This might seem surprising if one takes into account that the original file's information content, as calculated by our program using Shannon's formula, is over 25 million bits, which in turn is over 3 megabytes of information. This is not a violation of Shannon's law, but comes from the fact that our program, as well as our basis for calculating information entropy, was limited to treating each character separately. Sophisticated compression mechanisms may also take into account the correlations between the characters and build a Huffman tree where some codes correspond to single characters, while others correspond to pairs, triplets, etc. Such compression algorithms are more complicated, since these programs must decide which characters are worth treating together. As our program **Poe** shows, however, such an approach may be very useful. For example, the pair TH occurs 118 320 times in Shakespeare, more than the thirteenth most common letter M. Shannon's formula applied to the probability distribution of an optimal mixture of single characters, pairs, and other 'tuples' will usually show a lower information content as compared to the same formula applied to the single-character distribution.

10

The bridges of Königsberg

Graph theory

The ancient Prussian city of Königsberg was divided into its northern and southern parts by the river Pregel. (Königsberg is now called Kaliningrad and is a city in an isolated enclave belonging to Russia, surrounded by Poland, Lithuania, and the Baltic Sea.) There were two islands on this river, and they were connected with each other and with both parts of the city by bridges, as shown in Fig. 10.1. For a long time the residents of Königsberg had tried to plan a route within the city which would take them through each of the bridges exactly once. Leonhard Euler showed in one of his papers that this

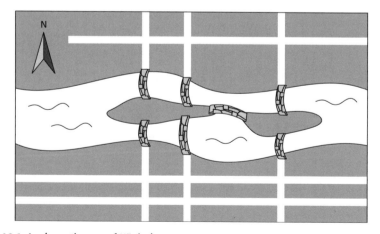

Fig. 10.1 A schematic map of Königsberg.

is impossible, and thus laid the foundations for what is now called **graph theory**.

Euler argued in the following way. Three bridges lead to the east island. Each time we cross one of these we either enter the island or leave it. Since three is an odd number, either the route begins on the east island and ends outside it, or vice versa. One way or another, one of the route's ending-points must be on the island. The same argument can be given regarding the west island, the north bank, and the south bank, since each of these sections of the city have an odd number of bridges leading to and from them. This would imply that a route which traverses each bridge exactly once would have to have at least four ending-points, which contradicts the fact that each route has exactly two such points—its beginning and its end.

Leonhard Euler (1707–1783); Swiss physicist, mathematician, and astronomer; author of a record-breaking number (886) of scientific papers.

The problem of the Königsberg bridges is a mere drop of water in the vast sea of problems and algorithms which make up a branch of mathematics called graph theory. The problem can be generalized to any configuration of bridges and islands. Besides questioning the existence of a path traversing all bridges, we may want to find the shortest such path, find out whether all parts of the city are reachable from a given point, color the map so that no adjacent regions are the same color, and so on.

Nodes and edges

Let us now consider the most important part of solving any problem—the choice of which model to use. In the case of the Königsberg bridges, our input data (in other words, our initial knowledge) is the map of Königsberg. But do we really need all of the information contained in this map to solve the problem? Do we need to know the name of the river, the names of the streets, or the name of the city itself? Of course, we do not. We do not even need to make the distinction between an island and a river bank, or care about the shapes of these objects and the distances between them. All we need to know is how the different regions of land are connected to each other by bridges. Mathematically speaking, we are interested in just the **topological properties** of the regions of land and the bridges.

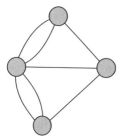

Fig. 10.2 A graph representing the bridges of Königsberg.

> The topological properties of objects remain unchanged under deform-
> ation, provided that the deformation does not break apart or join
> objects.

The data contained in the Königsberg map, which is relevant to solving the route problem, pertains to two types of objects, namely land areas and bridges. There are four land areas (the north bank, the south bank, the east island, and the west island) and seven bridges. The information about how these objects are connected is depicted in Fig. 10.2.

Figure 10.2 represents a **graph**. A distinction is often made between a graph and a diagram. A graph is an abstract mathematical structure, while a **diagram** is a picture—a visual representation of a graph. Graphs are composed of **nodes** and **edges**. In this case the nodes represent land areas and the edges represent bridges. Nodes which are connected by an edge are referred to as **neighbors**. A set of nodes connected by a set of edges is the minimal model of a graph. If required, this model can be enriched in a number of ways. We may associate labels (text) or weights (numbers) with the nodes and/or with the edges. We may also specify the direction of every edge, as if the bridges in our example were one-way streets. A graph whose edges have a specified direction is called a **directed graph**.

Trees, such as the ones produced by our **Huffman** program and described in Chapters 5 and 9, are special cases of graphs. By definition, a **tree** is an acyclic connected graph (see Fig. 10.3(a)). Acyclic means that it does not contain any closed paths. Most often, one of the nodes in a tree is distinguished as the root, which implies a direction to each of the edges—either toward the root node or away from it (see Fig. 10.3(b)). Each node in a tree, except for the root, has exactly one neighbor which is closer to the root than this node. That neighbor is called the **parent** of the node and all of the remaining neighbors are called its **children**. A node without children is called a **leaf** (see Fig. 10.3(b)). The Huffman tree has additional information associated with it. Each branch

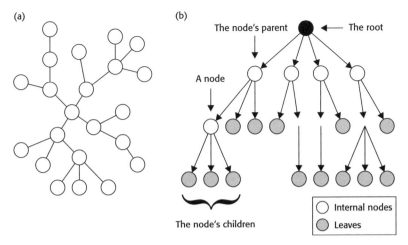

Fig. 10.3 Two graph diagrams: (a) a tree, and (b) the same tree with one node distinguished as the root and with the node relationships following from those marked.

is labeled either 'yes' or 'no', and each node has a weight associated with it which is proportional to the probability of reaching that node.

Graph problems

The problem of the Königsberg bridges is a special case of a whole class of problems. The aim was to decide whether any given graph contains an Euler path or an Euler circuit. A circuit is a closed path, one which begins and ends at the same node. By definition, an Euler path or Euler circuit traverses each edge exactly once. It turns out that a graph contains an Euler circuit if and only if it is connected and each of its nodes has an even number of neighbors. The condition for the existence of an Euler path is slightly weaker than that for the circuit, as it allows two nodes to have an odd number of neighbors (in such a case, the odd nodes must be the ones where the path begins and ends). A connected graph is one in which one can get from any node to any other node by following the edges of the graph. Knowing this rule makes it very easy to solve the Euler path and Euler circuit problems.

Let us now consider a slightly different problem. A shoe salesman lives in Oxford. Every day he visits six surrounding cities to deliver his merchandise. The problem is to find the shortest route which begins and ends in Oxford and which visits all of the cities on the salesman's agenda. The input data is a map of the vicinity of Oxford see Fig. 10.4. This data is a graph whose nodes are labeled with city names, and the edges have weights corresponding to the lengths of the roads connecting the cities. (These weights may not agree

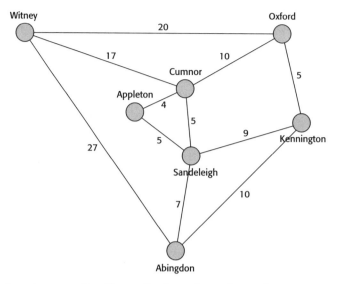

Fig. 10.4 A graph representing cities on the shoe salesman's agenda.

with the distances between cities, since roads are not straight lines.) Using the terminology of graph theory, we are looking for a **Hamilton circuit** in the graph with the lowest total weight. A Hamilton circuit is a circuit which visits each node in the graph exactly once. Note that, while in the Euler problem the stress was on the edges, the nodes are now playing the crucial role and in addition the edges carry weights.

Sir William Rowan Hamilton (1805–1865); Irish mathematician and astronomer; recognized as a genius from his childhood years. By the age of five he had already mastered Latin, Greek, and Hebrew, and at the age of twenty-two he was appointed a professor at Trinity College in Dublin.

A generic version of this problem, called the traveling salesman problem, is dreaded by computer scientists to this day. An algorithm which would solve the problem for any large graph in a reasonable time has not been found. The simple example of the graph above contains just two Hamilton circuits. Finding, measuring, and comparing these is not much of a problem. However, the number of such circuits in a graph can be, in the general case, proportional to the exponent of the number of nodes. All known algorithms for solving the traveling salesman problem are characterized by an exponential increase in time with the number of nodes in the graph. Such exponential algorithms are useless in practice, since the exponential function grows very rapidly. For a number of nodes between twenty and thirty we need an 'age of the universe' amount of time to solve the problem.

The traveling salesman problem belongs to a whole family of problems called **NP-complete** by computer scientists. An NP-complete problem is one for which an exponential algorithm can be easily found, but no algorithm exists which solves the problem in a shorter time. The amazing thing is that it has not been *proven* that better than exponential algorithms for NP-complete problems *do not* exist. Moreover, if one such algorithm could be found for just one of the problems in the family, then the algorithms for the remaining problems would follow automatically. Needless to say, finding such an algorithm, or the proof that it does not exist, is the dream of every computer scientist.

Observing many natural phenomena (one such example is the protein-folding problem) suggests that nature is able to 'solve' NP-complete problems quickly, which gives us hope that perhaps we will be able to do it as well some day. Perhaps the answer lies in computer models other than the ones we use now, like quantum computers or DNA computers.

It turns out that just deciding if a graph has a Hamilton circuit (or path) or not is NP-complete. In fact, graph theory, like many other fields of science, is full of NP-complete problems. Luckily, many of them can be solved in an approximate way—a way which finds a good solution rather than the very best one. For the Hamilton circuit problem, a fast (not exponential) algorithm has been found which does not always find the shortest path, but the path it finds is never more than twice as long as the truly shortest one. Such an imperfect solution must suffice when the known exact algorithms take millennia to complete.

 The program **Euler** attempts to find the shortest Hamilton path in a given graph by analyzing all possible node permutations. One can observe that with each additional node this task takes dramatically (exponentially) more time.

Another famous graph problem is that of coloring a graph so that no two neighboring nodes are of the same color, using the least possible number of different colors. It turns out that a planar graph, one which can be drawn on a plane without any edges crossing, can always be colored using at most four colors. This is a useful thing to know when painting political maps, if the neighboring countries are treated as neighboring nodes in a graph. A country neighborhood graph will always turn out to be planar. However, in the general case, determining the minimum number of colors needed to color a graph is NP-complete.

Graph applications

So far we have shown how graphs may be used to describe finite objects, such as a network of cities or the topology of bridges on a river. Graphs, however, are a very powerful tool for modeling infinite objects as well. To illustrate this, let us consider an infinite set of sentences defined as follows.

- A sentence begins with the words 'Jane has', contains a positive number of animal descriptions separated by 'and' conjunctions, and ends with a period.
- An animal description begins with the 'a' article, followed by any number (including zero) of 'very' adverbs, followed by one of three possible adjectives ('little', 'cute', or 'big') and one of two possible nouns ('dog' or 'cat').

In mathematical terms, a family of sentences is called a language. Examples of sentences belonging to the language defined above are as follows.

- Jane has a little dog.
- Jane has a very cute cat and a big dog.
- Jane has a very big dog and a very very big dog and a very little cat.

This family of sentences is a subset of the English language, since each of its members is a correctly constructed English sentence. It is also an infinite set, since sentences containing any number of 'very' words and any number of animal descriptions are allowed (even though only a finite number of *different* animal descriptions can be produced using this model).

For years, scientists have been struggling to program a computer to recognize, or even understand in some way, the written language. As we see from the above example, this endeavor is not an easy one, if only for the reason that the number of possible sentences is infinite. What is finite, however, is the set of rules governing the construction of sentences. The set of rules for our simple example may be represented by the graph depicted in Fig. 10.5.

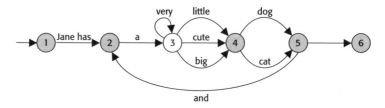

Fig. 10.5 A finite-state automaton defining a simple language.

The graph in Fig. 10.5 is a directed one, with a distinguished beginning and ending node and edges labeled with phrases. What the graph actually models is an abstract mathematical object called a **finite-state automaton**. Finite-state automata can be used to recognize sentences belonging to a language or to generate such sentences. The automaton in Fig. 10.5 models our simple 'Jane has...' language. Each path in the graph from the beginning node to the ending node corresponds to a sentence in the language. The labels of the edges along the path make up the sentence.

Even though the 'Jane has...' language is infinite, its model (the finite-state automaton) is finite. Moreover, the model may be easily understood and implemented on a computer. Languages which can be modeled using a finite-state automaton are called regular according to the classification given by Avram Chomsky. This classification includes a number of more complicated language families and their corresponding more complicated automata. While regular languages are at the bottom of this classification, recursive languages, modeled by the Turing machine described in Chapter 15, are at the top. Natural languages fall somewhere in the middle, though the debate on where exactly they belong is still going on. It seems, therefore, theoretically possible to teach a modern computer to 'speak' our language by constructing the appropriate model of this language. It must be the mere vastness of the rules governing natural language that has rendered the task inachievable to this day. Nevertheless, some subsets of our language can be modeled, even as simply as in the above example. Such models suffice to make computers understand the user's written input as long as that input conforms to relatively simple rules. Computer games, interactive help systems, and many other types of programs take advantage of this. Programming languages are also modeled by finite-state automata.

Avram Noam Chomsky (1928–); professor of linguistics at Massachusetts Institute of Technology. Owing to his knowledge of both fields, he applied rigorous mathematical descriptions to linguistic concepts.

Andrew C. Bulhak of Monash University has developed a system (the postmodernism generator) for generating essays using finite automata such as the one described above. His program randomly traverses predefined graphs (like the one in Fig. 10.5) to generate completely meaningless, but surprisingly realistic texts. The following is a sample excerpt.

'Culture is a legal fiction', says Foucault; however, according to Brophy [1], it is not so much culture that is a legal fiction, but rather the futility, and subsequent rubicon, of culture. But the subject is contextualised into a constructive capitalism that includes sexuality as a totality. Lyotard uses the term 'textual neodeconstructive theory' to denote a self-sufficient paradox.

Graphs play an important role in physics as well, the most developed example being **Feynman diagrams** in theoretical physics. A Feynman diagram is a highly-simplified model of a physical phenomenon involving elementary particles. Initially, Feynman diagrams described only processes involving photons and electrons, but the model was subsequently extended to accommodate the scattering and production of all elementary particles.

 Richard Phillips Feynman (1918–1988); American physicist. A vivid figure in the scientific world; known to the general public for his autobiographies.

Figure 10.6 illustrates two different Feynman diagrams. Figure 10.6(a) models the collision of two electrons (represented by straight-line segments). The electrons interact in the process by exchanging two photons (wavy-line segments). Figure 10.6(b), in turn, represents the collision of two photons, a process which results in the production of an electron–antielectron pair. In addition, the produced electron interacts with the electromagnetic field by emitting and subsequently absorbing a photon. Feynman diagrams are partly directed graphs, since they may contain both directed and undirected edges. The directed edges, indicated by line segments with arrows, signify moving charges. Since the electron and antielectron in Fig. 10.6(b) have opposite charge, their corresponding edges are oppositely directed.

As the example shows, Feynman diagrams are a handy means of depicting physical processes. However, if their role was solely that, then they would not have earned Feynman a Nobel prize. The point is that Feynman diagrams allow

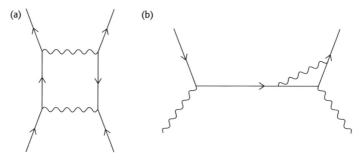

Fig. 10.6 Examples of Feynman diagrams, which represent (a) the collision of two electrons, and (b) the collision of two photons.

us to conduct precise calculations. Feynman assigned a specific physical quantity to each diagram element (node, and wavy- and straight-line segments), which in turn allows us to calculate the probability that the process modeled by a particular diagram will occur. Nowadays, and with the help of computers, calculations based on Feynman diagrams have been brought to perfection. Some processes investigated by theoretical physicists took as many as several hundred Feynman diagrams to model.

Toward the end of the twentieth century, the American mathematicians Duncan Watts and Steve Strogatz investigated a different kind of graph, namely the graph of social connections. This graph contains a node for every person in the world and an edge between nodes if the corresponding people know each other. Naturally, it would take a gigantic effort to actually plot this entire graph, since the information contained in it is huge. Nevertheless, some traits of this network have been observed indirectly—most notably the 'six degrees of separation' property.

'Six degrees of separation' within the social network was discovered in 1967 by the American social psychologist Stanley Milgram, who sent letters to a random selection of people asking them to forward the letter to a specified friend of his. Should they not know this person, they were to forward the letter to someone they considered to be more closely related to him. Most of Milgram's letters reached his friend after having been forwarded approximately six times. This led to the formulation of the **small world hypothesis**, namely that any two people in the world are connected by a path of approximately six social connections. Milgram's experiment was repeated on a much larger scale using electronic mail, and the results confirmed Milgram's findings. A second edition of this experiment, called the *Small world project*, is ongoing (see http://smallworld.columbia.edu for more information).

Watts and Strogatz found that, even though the social connections graph is, to a degree, random, many general observations regarding its structure can be made. What is more, it turned out that the social network is not the only type of graph observed in nature which has such a structure. Power lines, Internet connections, but also neurons in a brain, food webs (what eats what in an ecosystem), and many others are like this as well. It seems that there is a special type of graph architecture which is intrinsic to most natural networks. The investigation of this architecture is ongoing, but it is already apparent that this model models surprisingly many (often seemingly unrelated) aspects of reality.

Prisoner's dilemma

Game theory

You are under arrest for having committed a crime together with another felon. The prosecutor offers you a choice of confessing to the crime or not, stating the following circumstances. If neither you nor your partner confesses, then you will get just one year of prison each, since the prosecution is short of evidence. If you both confess, then you will be sentenced to ten years of prison each. However, if one of you confesses and the other claims his innocence, then the one who confessed will be set free in return for testifying against the other, while the other will get twenty years. You are aware that your partner in crime was informed of the same conditions and has the same choice, but you have no way of guessing his decision or communicating with him in any way. You do not care what happens to your partner, you just want to get the lowest sentence yourself. What should you do?

The problem of choosing between confessing to the crime and claiming innocence in the situation described above is called the **prisoner's dilemma**. The American mathematician Albert W. Tucker was the first to introduce it in 1950, during a lecture at Stanford University, while he aimed to illustrate the diffculties arising during the analysis of some games. Since then, the prisoner's dilemma has found numerous applications in such diverse fields as psychology, sociology, economy, warfare, ecology, and others.

The theory of games

The branch of mathematics dealing with problems such as the prisoner's dilemma is called **game theory**. Game theory in general provides mathematical models for making decisions in a conflict situation. The conflict in question may come simply from the opposing interests of parties involved in a game, but when modeling war strategies it is a conflict in the full meaning of the word. As we have noted earlier, game theory does not apply merely to pastimes—its main significance is in the areas of economy and warfare.

The aim of game theory is to provide an optimal strategy in a conflict situation. It is enough to glance at the history and artwork of mankind to realize how big a part various types of conflicts play in our lives. Most of them cannot be described by models as simple as the ones presented below. Still, the simple examples described in this chapter illustrate how conflict situations may be analyzed mathematically and how an optimal strategy may be chosen rationally.

 Oskar Morgenstern (1902–1977); Austrian economist. Immigrated to the United States in 1938, where he became a professor at Princeton University.

Game theory in its present form came into existence in 1928, when John von Neumann published a mathematical proof of the minimax theorem—a fundamental theorem of game theory. For almost twenty years, only mathematicians took an interest in the development of game theory. In 1944, however, von Neumann co-authored the book *Theory of games and economic behavior* with the known economist Oskar Morgenstern. It was then that game theory found its way to a wider audience and to its well-deserved key position in the modeling reality domain.

> A procedure can be called a game only when its participants, the players, have a chance of making decisions.

Most often, players make a number of decisions at different stages of the game. In order to standardize the model of game playing, we will consider this whole series of decisions as a single choice—the choice of a **strategy**. A strategy defines what decision is to be made given any possible game situation. The prisoner's dilemma has just two possible strategies, since each player makes exactly one decision, always in the same situation. One strategy is to claim innocence and the other is to confess to the crime. For most actual games, however, the set of possible strategies is much too large to be enumerated and can only be analyzed in a limited way. This is what makes games fun. Even such a simple game as *tic-tac-toe* (known in the United Kingdom as *noughts and crosses*) has

		Prisoner I	
		Deny	Confess
Prisoner II	Deny	I:1 II:1	I:0 II:20
	Confess	I:20 II:0	I:10 II:10

Fig. 11.1 The payoff matrix for the prisoner's dilemma.

thousands of possible strategies. The number of essentially different *tic-tac-toe* games is 26 830 (see http://www.adit.co.uk/html/noughts and _crosses.html). This number can be estimated as follows. The first player has a choice of nine places as to where to put his mark, the opponent has eight, and so on, which at first seems to result in $9! = 362\,880$ possible games. However, the symmetry of the board under rotations and reflections leads to a significant reduction. Moreover, some of the games end while there are still empty slots. While *tic-tac-toe* is still a game for which all of the strategies can be feasibly listed, for more elaborate games (for examples, *chess, checkers, and Go*) this cannot be done. Supposing that the average number of moves per *chess* match is thirty for each player and that each move is a choice between an average of at least twenty possibilities, then there are something like 20^{60} possible *chess* games. Even if most of them can be easily recognized as unreasonable, an unimaginably large number of strategies are still left to be analyzed.

Analyzing strategies (when they can all be listed) can be aided by the use of a table called the **payoff matrix**. A payoff matrix for n players is n-dimensional. There is an entry for every possible combination of the players' strategies. It represents the result of the game in the case when the players chose these strategies. The payoff matrices of two-player games are two-dimensional, and hence can be depicted on a piece of paper. In the case of *chess*, the paper would be, by several orders of magnitude, too large to fit on the surface of the Earth; but a payoff matrix for the prisoner's dilemma is just two by two (see Fig. 11.1). Each prisoner has just two strategies to choose from, namely confession or denial of the crime. The payoff matrix specifies the score of each player in the case of every possible strategy combination. In this case the payoffs correspond to years spent in prison, so a lower value is better than a higher one.

Betrayal and cooperation

The prisoner's dilemma possesses a **dominant strategy**. A dominant strategy is one that yields the most advantageous result for every possible strategy chosen by the opponent. If the opponent chooses to confess, then our better strategy is to confess as well, since we get ten years instead of twenty. If our opponent chooses to deny committing the crime, then again our better strategy is to confess, since we get out free instead of getting one year. Since confession is the better choice in all of the cases, it is the dominant strategy.

Following these arguments, it seems reasonable to confess to the crime rather than not to. But let us look at the problem from a different perspective. If both you and your partner reason in terms of dominant strategies, then you will both choose confession and consequently both get ten-year sentences. Meanwhile, had you both claimed your innocence, then you would have received just one year each—an outcome which is much better for both players! This is the whole point of the dilemma. The lack of trust between the parties may lead to them choosing an option which is worse for both.

The prisoner's dilemma models many real-life situations. A classic example is an arms race; two countries have a pact about not increasing their weaponry, but each country takes into account the possibility of the other secretly breaking this pact. Each country finds its dominant strategy by reasoning along the following lines. If the other country secretly increased its military power, then we would be better off if we had done the same, otherwise we may get invaded and conquered. In the case of the other country being loyal to the pact, it is still useful to increase the weaponry—and perhaps invade the other country. Consequently, both countries get more weapons.

The problem with choosing the right strategy is that we have no information about what the other party is going to do. We have no reason to trust the other party to choose the option which is good for both, so we choose the option which is bad for both in order to avoid a situation in which the opponent gets the best and we get the worst. The problem becomes really interesting, however, if we allow the two parties to play the game over and over again, and base their decisions on the outcomes of the previous encounters. This enhanced variation is called the **iterated prisoner's dilemma**. Unlike the case of the single game, the iterated prisoner's dilemma strategies are not easy to analyze. If a player makes his decision on the basis of the outcomes of the preceding three encounters, then this still yields 2^{64} possible strategies, since there are $4^3 = 64$ possible combinations of three game outcomes. It is diffcult, if not impossible, to establish which strategy is generally best for the iterated prisoner's dilemma, but some heuristic rules can be found by various means.

The most successful method of analyzing the iterated prisoner's dilemma strategies which has been put forward so far was presented by Robert Axelrod. In his 1984 book *The evolution of cooperation*, Axelrod described a genetic algorithm designed to find an optimal strategy for the problem. Genetic algorithms, as well as their application in this particular case, will be described in detail in Chapter 12. At this point, let us just say that Axelrod's algorithm generally favors the tit-for-tat strategy and always claiming innocence at the beginning of the game, though the exact results depend on the set-up of the genetic algorithm. Tit-for-tat means that you should do what your opponent did in the previous move, but in this case it is definitely an over-simplification. Axelrod's algorithm speaks in favor of such things as forgiveness—claiming innocence in the case of a detected 'arms race' (a series of both parties claiming guilty) in the hope that the opponent will detect this sign of good will and will revert to the strategy which is optimal for both. In general, a simulated society of iterated prisoner's dilemma players tends to evolve toward cooperating with each other for their mutual benefit.

 Robert Axelrod (1943–); political science professor at the University of Michigan with a BA in mathematics; best known for his work on the evolution of cooperation which has been cited in over three thousand articles.

 The program **Axelrod** implements a genetic algorithm for evolving strategies for the iterated prisoner's dilemma. The program simulates a population of prisoners, each of whose genetic code describes a strategy. The program intermittently arranges tournaments of iterated prisoner's dilemma within the population and breeds its members, favoring the ones who obtained a better net score in the tournament. The mechanisms of evolution cause the population to develop beneficial playing strategies. Among other things, the user may observe which decision, deny or confess, is more favored in a given situation. A situation in this case refers to the combined outcome of the last three encounters. It is important to note, however, that the simulated organisms develop strategies by playing against each other. Therefore, whatever strategy turns out to be optimal, it is optimal only in the environment composed exclusively of equally-'evolved' organisms. An optimal strategy for playing against, say, a random opponent could turn out very different.

The difficulty in finding a solution to the prisoner's dilemma is largely a consequence of the fact that the two parties are not exactly playing *against* each other, but are also not exactly playing *with* each other. One player's gain is not necessarily the other's loss. Games with a balance of gains and losses are called **zero-sum games** and are much easier to analyze. Zero-sum games have a constant net payoff in each field of the payoff matrix. It is only logical to set this constant to be zero, and hence the name. In other words, what one player gains is what another loses. The vast majority of actual games have this property. Because of their symmetry, for zero-sum two-player games it is enough to list only one of the player's payoffs in the fields of the matrix, since the other player's payoffs will always be simply opposite numbers. What von Neumann proved was that every zero-sum game has an optimal strategy.

Finding the optimal strategy

We will illustrate the concepts related to zero-sum games using a game show scenario. In this example, the game show participants, represented by a family of four people, play against the game show organizers, represented by the game host. The stake is a brand new car. This is a zero-sum game since exactly one side (the participants or the organizers) is going to keep the car in the end, no matter what strategies they choose.

The All stars game show is based on answering questions from the four categories Movie stars, Pop stars, Sports stars, and TV stars. One family has just made it to the final stage of the game, its members being Dad, Mom, Boy, and Girl. At this point the family is supposed to choose one person among themselves to answer the final question. If the elected representative answers the question correctly, then the family will win the brand new car. If he or she answers incorrectly, then the brand new car will remain the property of the television network. The family needs to choose a representative in order to maximize their chances of winning, but without knowing what category the final question will belong to. In the meantime, the game show host has to solve an analogous problem: to choose the final category so as to minimize the chances of the participants winning the car, without knowing whom the family will designate to partake in the finals. Both parties have the same statistical data to base their decisions on, the skill of each family member in each game category, as obtained from the preceding stages of the game. This data is summarized in Fig. 11.2, where each entry is the chance, in per cent, of the given family member correctly answering a question from the given category.

The two competing parties, the family and the host, reason along the lines outlined below.

- **The family**
    —If we choose Dad and if the host chooses Dad's worst category, Movie, then we have a 30% chance of winning the car.
    —If we choose Mom and if the host chooses Mom's worst category, Sports, then we have a 30% chance of winning the car.
    —If we choose Boy and if the host chooses Boy's worst category, Pop, then we have a 60% chance of winning the car.
    —If we choose Girl and if the host chooses Girls's worst category, Pop, then we have a 20% chance of winning the car.
    —We choose Boy! It is our best choice.

	Movie	Pop	Sports	TV	α_i
Dad	30	50	90	80	30
Mom	80	40	30	70	30
Boy	70	60	80	90	(60)
Girl	70	20	40	60	20
β_j	80	(60)	90	90	

Fig. 11.2 Family members' skills in different categories.

- **The host**
  - If I choose Movie and if the family chooses Mom, who is most skilled in the Movie category, then I have an 80% chance of losing the car.
  - If I choose Pop and if the family chooses Boy, who is most skilled in the Pop category, then I have a 60% chance of losing the car.
  - If I choose Sports and if the family chooses Dad, who is most skilled in the Sports category, then I have a 90% chance of losing the car.
  - If I choose TV and if the family chooses Boy, who is most skilled in the TV category, then I have a 90% chance of losing the car.
  - I choose Pop! It is my best choice.

The skills table in Fig. 11.2 has been augmented with a column labeled α_i. The value of α_i is the worst-case-scenario game result, should the family choose the ith family member. The value α_i is the *minimum value* in the ith row. Analogously, β_j is the worst-case-scenario game result, from the host's perspective, should he choose the jth category. The value β_j is the *maximum value* in the jth row. The family chooses the member with the highest α_i, while the host chooses the category with the lowest β_j. In this way, both parties optimize their result in the worst-case scenario. Since we choose the maximum value out of minimum values and the minimum value out of maximum values, this method is called the **minimax principle**.

It is important to note that, in this case, the choices made by the family and by the host are both optimal. If the family picks Boy, then the host, choosing anything other than Pop, decreases his chances of keeping the car. Likewise, if the host picks Pop, then the family, choosing anyone other than

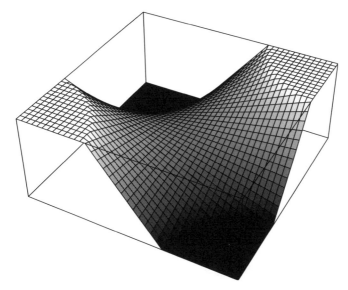

Fig. 11.3 A saddle formed by the plot of the function $f(x, y) = xy$.

Boy, decreases their chances of winning the car. Such a mutually-optimal solution exists because the payoff matrix (the skills table in Fig. 11.2) contains a **saddle point**. A saddle point is an element of a matrix with a value lowest in its row and highest in its column. The name obviously comes from the shape of a saddle—an object which curves down toward the sides and up to the front and back, much like the plot of the function $f(x, y) = xy$ depicted in Fig. 11.3. A real saddle (see Fig. 11.4) has a point which is simultaneously the lowest along one axis and the highest along the other.

If a game's payoff matrix contains a saddle point, then both players' strategies corresponding to that point are optimal. Moreover, the two strategies are **pure strategies**, meaning that they define the player's choice unambiguously. The alternative to a pure strategy is a mixed strategy, a concept to be defined shortly. If a game's payoff matrix contains more than one saddle point, then its solution corresponds to the most profitable one. If it has more than one equivalent (in terms of gain) saddle point, then the game has multiple solutions.

The minimax principle described above can be formulated formally. Let us denote the elements of the payoff matrix as a_{ij}, where i denotes the row and j the column of the matrix. In general, the matrix can be rectangular, and not necessarily square. The quantities α_i and β_j, as well as α and β, may be defined as follows:

$$\alpha_i = \min_j a_{ij}, \quad \beta_j = \max_i a_{ij}, \quad \alpha = \max_i \alpha_i, \quad \beta = \min_j \beta_j.$$

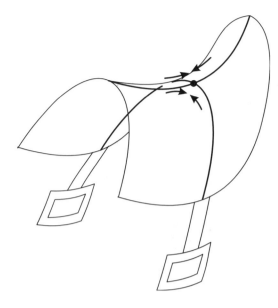

Fig. 11.4 The saddle point of an actual saddle is the lowest point in the lengthwise cross-section of the saddle going through that point. Simultaneously, it is the highest point in the breadthwise cross-section of the saddle going through that point.

The numbers α and β are called the **lower game limit** and **upper game limit**, respectively. In brief, they can be defined in the following way:

$$\alpha = \max_i \min_j a_{ij}, \quad \beta = \min_j \max_i a_{ij}.$$

It is clear how the above equations give rise to the term **minimax**.

The payoff matrix in Fig. 11.2 does not give rise to very interesting results. It is immediately obvious which strategy is optimal for each of the players, and any divergence from that strategy gives a worse result. This is because the two game limits α and β are equal. However, usually this is not the case. Let us change the payoff matrix to be that shown in Fig. 11.5. If the two parties were to reason along the same lines as before, then this time the family would choose Mom (a decision guaranteeing a 40% chance of winning the car) and the host would choose Pop (a decision yielding a 50% chance of losing the car in the worst case). This time the lower limit is smaller than the upper one: $\alpha = 40\% < \beta = 50\%$. The payoff matrix does not contain a saddle point. Can the family, therefore, obtain a better result? This question is answered by von Neumann's minimax theorem, which is stated as follows.

Every zero-sum two-person game has an optimal strategy, but usually the strategy is mixed.

	Movie	Pop	Sports	TV	α_i
Dad	30	50	90	80	30
Mom	80	40	80	90	(40)
Boy	30	40	60	50	30
Girl	20	30	70	70	20
β_j	80	(50)	90	90	

Fig. 11.5 A different version of the family members' skills table.

A **mixed strategy** is one which incorporates elements of risk and bluff. The new payoff matrix, with different α and β values, opens up ways for both sides to improve their strategy. The two opposing parties may now think along the following lines.

- **The family's first reasoning**
  The host will choose Pop, since it guarantees him the lowest chance of losing the car. Taking that into account, we should elect Dad and obtain a 50% chance of winning.
- **The host's first reasoning**
  The family thinks that I will choose Pop, since it guarantees me the lowest chance of losing the car, so they will elect Dad. Therefore, I will choose Movie and reduce their chances of winning to 30%.
- **The family's second reasoning**
  The host thinks that we will elect Dad, counting on his safe choice of Pop, and so he will chose Movie. So if we choose Mom, then we will have an 80% chance of winning.
- **The host's second reasoning**
  The family must realize that I have seen through their plan to elect Dad, counting on my safe choice. In that case they will elect Mom to optimize their chance of winning. Knowing that, I should choose Pop and minimize their chance of winning.
- **And so on...**

The above illustrates that neither of the sides has an *obvious* strategy. Departing from the 'safe' strategy (one with the most beneficial worst-case scenario) may lead to an improved score if the opponent persists with *his* safe strategy. However, it also presents a risk of decreasing one's score, should the opponent foresee this decision. Both parties reasoning in this way leads to a vicious circle of 'he thinks that I think that he thinks that I think that he thinks...'. The only way to overcome this is to select a strategy at random. Only this can prevent the opponent from anticipating the decision. This leads us to conclude that the best solution for each of the sides is a random choice of one of the options, perhaps with some uneven probability distribution.

Probability is hard to define with respect to a single event. In order to properly explain the concept of a mixed strategy, we need to imagine a whole series of similar events. In this case, let us imagine that the family and the host get to make their choice a repeated number of times, each time according to the same payoff matrix shown in Fig. 11.5. We have already established that each time the family should elect a player at random (so as not to be predictable to the host), with a certain probability associated with each family member. Similarly, the host chooses a random category with different probabilities associated with each category. The optimal mixed strategy for a given player is defined by the set of probabilities, which gives the best result *on average*.

A mixed strategy for the family is defined by the set of four probabilities $\{p_1, p_2, p_3, p_4\}$ of electing Dad, Mom, Boy, and Girl, respectively. Likewise, the host's mixed strategy is the set of four probabilities $\{q_1, q_2, q_3, q_4\}$ assigned to the categories Movie, Pop, Sports, and TV. Given these probabilities, we may calculate the average outcome of the game, in the case for which the host chooses category number j:

$$\langle \alpha_j \rangle = p_1 a_{1j} + p_2 a_{2j} + p_3 a_{3j} + p_4 a_{4j}.$$

Analogously, we may calculate the average game outcome, in the case for which the family elects family member number i:

$$\langle \beta_j \rangle = q_1 a_{i1} + q_2 a_{i2} + q_3 a_{i3} + q_4 a_{i4}.$$

We will have found an optimal solution for each of the players if we can find a set of probabilities p_i such that the values of $\langle \alpha_j \rangle$ are equal for all j and a set of probabilities q_j such that the values of $\langle \beta_i \rangle$ are equal for all i (i.e. $\langle \alpha \rangle = \langle \alpha_1 \rangle = \langle \alpha_2 \rangle = \langle \alpha_3 \rangle = \langle \alpha_4 \rangle$ and $\langle \beta \rangle = \langle \beta_1 \rangle = \langle \beta_2 \rangle = \langle \beta_3 \rangle = \langle \beta_4 \rangle$). In this case the family's average score will not depend on the host's choice of category and the host's average score will not depend on the family's choice of family member. The minimax theorem states that such a set of probabilities always exists and, moreover, that both optimal outcomes, $\langle \alpha \rangle$ and $\langle \beta \rangle$, are equal. This common optimal outcome is called the game's **value**. If we denote the game value by v, then $\langle \alpha \rangle \langle \beta \rangle = v$.

With this knowledge, we may proceed to find the optimal strategies in the case of the *All stars* game show example. To make the task simpler, let us first make the following observations. The host's choices of Sports or TV are completely irrational, since a choice of Movie or Pop is always better for him than either of the other two, *regardless* of whom the family elects. In game theory terms, we would say that Movie and Pop dominate Sports and TV, and we may disregard the Sports and TV categories altogether. A similar dominance of Mom and Dad over Boy and Girl from the family's point of view leads to the conclusion that we may disregard the election of either of the children, since they are irrational choices for the family. This gives $p_3 = p_4 = q_3 = q_4 = 0$, and the mixed strategy comes down to a choice between Mom and Dad and between Pop and Movie. Notice that, during our speculations about the family's and the host's intermittent reasoning, the family's choice also oscillated only between Mom and Dad, while the host was trying to make his mind up between Pop and Movie. The other options were never taken into account, since one of the dominant strategies was always better.

With only dominant strategies left, the payoff matrix has shrunk to the two by two table depicted in Fig. 11.6. In general, the formulae for optimal mixed strategies in the case of a two by two payoff matrix are as follows:

$$p_1 a_{11} + p_2 a_{21} = v, \quad q_1 a_{11} + q_2 a_{12} = v,$$
$$p_1 a_{12} + p_2 a_{22} = v, \quad q_1 a_{21} + q_2 a_{22} = v,$$
$$p_1 + p_2 = 1, \quad q_1 + q_2 = 1.$$

Such a linear set of equations can be easily solved by eliminating each of the unknowns one by one. In theory, a set of equations with more equations than unknowns (as is the case here) could turn out to be unsolvable, but the minimax theory guarantees that this will not happen. The solution in the

	Movie	Pop
Dad	30	50
Mom	80	40

Fig. 11.6 The payoff matrix with only dominant strategies left.

general case is

$$p_1 = \frac{a_{22} - a_{21}}{a_{11} + a_{22} - a_{12} - a_{21}}, \quad q_1 = \frac{a_{22} - a_{12}}{a_{11} + a_{22} - a_{12} - a_{21}},$$

$$p_2 = \frac{a_{11} - a_{12}}{a_{11} + a_{22} - a_{12} - a_{21}}, \quad q_2 = \frac{a_{11} - a_{21}}{a_{11} + a_{22} - a_{12} - a_{21}},$$

$$v = \frac{a_{11}a_{22} - a_{12}a_{21}}{a_{11} + a_{22} - a_{12} - a_{21}}.$$

Substituting the actual values from Fig. 11.6 for a_{ij}, we obtain the following solution:

$$p_1 = \frac{2}{3}, \quad p_2 = \frac{1}{3}, \quad q_1 = \frac{1}{6}, \quad q_2 = \frac{5}{6}, \quad v = 46\frac{2}{3}.$$

These results mean that the best strategy for the participants of the *All stars* game show is throwing a dice. The family should elect Mom if the dice shows 1 or 2, and they should elect Dad if the dice shows 3, 4, 5, or 6. In this way they will achieve the desired probability distribution of $p_1 = 2/3$ and $p_2 = 1/3$. Likewise, the host should throw a dice, choosing Movie only if the dice shows 1 and Pop otherwise, to achieve his probability distribution of $q_1 = 1/6$ and $q_2 = 5/6$. The two mixed strategies give, on average, a $46\frac{2}{3}\%$ chance of winning the car, and neither of the players can improve his strategy to obtain a more beneficial result for himself.

Beyond minimax

The minimax theorem only applies to zero-sum two-player games. Games which involve more players or those without a gain/loss balance very often do not have such straightforward solutions, as did the *All stars* game in our example. They might, however, have a number of **Nash equilibria**.

 John Forbes Nash (1928–); American mathematician who, despite his battle with schizophrenia, contributed greatly to game theory and differential geometry. He is widely known from *A beautiful mind* (2001), an Oscar-winning movie based on his biography.

The concept now referred to as Nash equilibrium was first introduced by John Forbes Nash when he was still a graduate student, and it earned its author the Nobel prize in economics many years later, in 1994. Optimal saddle points for games which have saddle points and optimal mixed strategies in the case of zero-sum two-player games are examples of Nash equilibria. The concept, however, extends further and applies to all kinds of games. A Nash equilibrium is an assignment of a strategy to each player such that no player can improve his average score by changing his strategy, assuming that all of the remaining players keep their strategies unchanged.

The only Nash equilibrium for the prisoner's dilemma is both players pleading guilty. If both players plead innocent, then one of the players can decrease his oneyear jail sentence by pleading guilty (testifying against the other) and coming out free. If one of the players pleads guilty and the other pleads innocent, then the one who claims innocence can improve his strategy by admitting to the crime and getting ten years instead of twenty. If both players claim guilty, however, then either player changing his strategy to innocent would increase his jail sentence from ten to twenty years.

If a game has one Nash equilibrium, then the logical reasoning of all the players will always lead to that set of strategies, just as the logical reasoning at the beginning of this chapter led to the guilty/guilty 'solution' of the prisoner's dilemma. Nash showed that every **finite game** (allowing only a finite set of choices) has at least one Nash equilibrium composed of mixed strategies. The implications of this fact stretch far beyond the scope of playing traditional games. Whether they are aware of it or not, all parties playing a part in what is called 'global economy' strive to find the Nash equilibrium of that overwhelmingly complex 'game'.

12

Let the best man win

Genetic algorithms

Undoubtedly the most fascinating aspect of our planet is the life with which it abounds. To this day, scientists have not been able to fully comprehend the mechanisms which gave rise to the remarkable richness and complexity of the life-forms around us. Nevertheless, a universally-accepted theory exists, which describes how life on Earth, as we know it, has come to exist. This theory is called the **theory of evolution** and was put forward in the nineteenth century by Charles Darwin. It explains how very simple natural mechanisms applied to generation after generation cause organisms to gradually develop and become better and better adapted to the environment in which they live.

Charles Darwin (1809–1882); British biologist; first to formulate the theory of evolution in his work The origin of species.

Natural **evolution** is a process which affects a population of living organisms. The mechanisms of evolution fall into two categories, namely ones which increase the diversity of the population and ones which selectively decrease it. The mechanisms belonging to the first category, such as **mutation** and **recombination**, make random changes to individuals belonging to the population. The mechanisms belonging to the second category, such as **natural selection**, remove from the population individuals who are worse adapted to the environment. Alternatively, they decrease the chances of the worse-adapted individuals reproducing.

> The long-term operation of evolutionary mechanisms causes the population's gradual improvement. This process is called evolution.

Evolutionary and genetic algorithms

The mechanisms of evolution which affect living creatures are natural phenomena. Nothing stands in the way, however, of applying these mechanisms artificially to collections of lifeless objects in order to achieve their improvement (an improved object is defined here as an object which better serves our purpose). Such a procedure is called an **evolutionary algorithm**. In general, evolutionary algorithms may be applied to anything and for any purpose. For example, such methods have been successfully used to produce complicated electronic circuits. First, a goal is set—the circuit must perform a certain task. Next, a random collection of circuits is produced. Subsequently, the circuits are used to perform the designated task. The ones which did a relatively better job are kept and the rest are discarded. The diminished 'population' of circuits is then 'bred'—its members are reproduced with the introduction of slight random changes. The testing, discarding, and reproduction are repeated many times. Of course, not all problems can be solved using this approach, but the above procedure has, in some cases, yielded effective solutions which have eluded human designers.

In order for evolution to take place, the population must be able to reproduce—to produce new members which resemble its old members. The reproduction of life on Earth is based on **genetic code**. Every living organism on Earth possesses a collection of macromolecules, which describe its form and function. This description, called the genetic code, is copied when an organism procreates, allowing offspring to resemble their parents.

Evolutionary algorithms which model genetic code to ensure that subsequent generations are improved versions of the previous ones are called **genetic algorithms**. Before we go on to describe genetic algorithms in detail, note that such algorithms are very simplified models of the genetic reproduction and evolution taking place on Earth—they mimic just the key features of the real process. The notions used in designing genetic algorithms often correspond to those used in biology, but their actual implementations are usually simplified variants of the original ones.

An algorithm is a procedure which acts on a set of data structures. The data structures which genetic algorithms operate on are summarized below.

- **Individual**—an entity possessing traits which distinguish it from other individuals and a set of genes encoding these traits; we assume that (unlike in real life) all of the individual's traits are stored in his genetic code.
- **Genotype**—a sequence of symbols constituting the genetic code of an individual, encoding his set of traits.
- **Phenotype**—an individual's set of traits, as encoded in the genotype.
- **Population**—a set of individuals.

The first stage of designing a genetic algorithm consists of giving answers to the following questions.

- What problem do we want to solve?
- What set of potential solutions to this problem do we want to search? (What phenotypes will be allowed?)
- How do we encode a potential solution as a sequence of symbols? (How do we define the correspondence of genotypes and phenotypes?)
- How do we decide which solutions (phenotypes) are better than others?

A genetic algorithm, in its most general form, is outlined in Fig. 12.1. The **recombination** (also called **crossing over**) of a set of genotypes is the

---

- Set up a population P of p individuals with random geno- types. (Set the generation counter to zero.)
- Repeat in a loop
  - Set up a new empty population N.
  - Repeat p times
    - Choose a random subset of the current population, **favoring** individuals with better phenotypes.
    - **Recombine** the genotypes of the individuals in the subset to create a new genotype.
    - **Mutate** the new genotype.
    - Add an individual with the new genotype to the new population.
  - Discard population P and set $P \leftarrow N$. (Increase the generation counter by one.)

Fig. 12.1 Schematic description of a genetic algorithm.

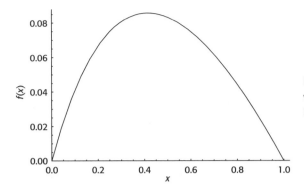

Fig. 12.2 The function whose maximum is sought by the genetic algorithm.

production of a new genotype whose different genes are taken from different members of the set. Recombination takes place in real life, when the parents' chromosomes cross over to form the genotype of the child, and effectively the child gets some genes from the father and some from the mother. In nature we can observe sets of two parents (sexual reproduction) or sets of one parent (asexual reproduction). In the case of asexual reproduction recombination is trivial—it is just the exact copying of the genes. Sets of three or more parents do not occur in real life, which leads us to believe that they would probably not be useful to implement in genetic algorithms either. **Mutation** is the introduction of random changes in a genotype.

A simple optimization problem

Now that we have outlined the general rules, let us proceed to the design of an actual genetic algorithm. The first step is answering the questions listed above.

- *What problem do we want to solve?*
  Find the maximum of the following function, shown in Fig. 12.2:

$$f(x) = 1 - \frac{1}{x+1} - \frac{x}{2}, \quad x \in [0, 1].$$

Genetic algorithms are, by their nature, suited for solving **optimization problems**. Optimization problems are those for which potential solutions can be compared and classified as *better* or *worse*. The *best* potential solution, according to this classification, is the actual solution to the problem. For example, the traveling salesman problem described in Chapter 10 is an optimization problem since it involves finding the shortest cycle in a graph. Potential solutions, i.e. cycles in the graph, can be compared according to their length (the shorter the better), and the best one is the solution to the problem. In general, optimization problems are those which seek the extremum (maximum or minimum) of a function, namely they find the x for which $f(x)$ is highest (or lowest). In the simplest case, x and $f(x)$ are numbers, but in general they can be very complicated objects. Some even use this approach to describe the process of evolution and the present distribution of species on Earth. In this case x is an organism and $f(x)$ is a measure of how well x is adapted to the environment, referred to as the organism's fitness. Organisms which are more fit are more likely to pass their genes on. The function $f(x)$ is customarily called the **fitness function**. When applied to real life, the fitness function is just a model, since an organism's fitness cannot be measured explicitly. The function is implicitly invoked by the observation that some organisms reproduce more successfully than others.

- *What set of potential solutions to this problem do we want to search?*
  We need to search the set of real numbers between 0 and 1. In theory this set is infinite. With a computer, however, we can only represent real numbers with a limited accuracy. Taking this into account, we search a finite set of real numbers which are evenly distributed within the [0, 1] range. The more numbers we search, the more accurate the solution. In this example we search 2^{32} possible numbers. The justification of this particular quantity follows from the succeeding paragraph.

- *How do we encode a potential solution as a sequence of symbols?*
  Since we are using a computer, the most logical alphabet to use for genes is the binary one. To encode one out of 2^{32} possible numbers we need exactly thirty-two bits. Binary notation is a way of encoding natural numbers from the range $[0, 2^{32} - 1]$ using thirty-two bits. All we have to do is combine this coding with a simple scaling function (dividing by $2^{32} - 1$) to obtain the desired range of values. In other words, the genotype is a sequence of thirty-two bits and the phenotype is its corresponding binary number divided by $2^{32} - 1$ (see Fig. 12.3). Later in this chapter we will discuss why this coding, though probably the simplest, is not the best choice for optimizing continuous real functions. We choose thirty-two-bit accuracy since it is convenient to implement (thirty-two-bit data storage is intrinsic to the platform we are using), though from the genetic algorithm point of view one can use any number of bits.

- *How do we decide which solutions (phenotypes) are better than others?*
  Since we are seeking the maximum of f, we will favor individuals whose phenotypes, when used as arguments of f, yield a larger number.

The simplest genetic algorithms are those which operate on a population of one member. Even such minimal genetic algorithms may prove sufficient

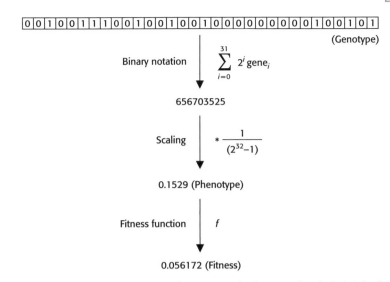

Fig. 12.3 Translating a genotype into a phenotype and subsequently calculating the fitness of the individual.

for solving an easy problem. With just one member, sexual reproduction is out of the question, which in turn eliminates the problem of implementing recombination. The remaining mechanism for scanning the solution space is mutation. A good implementation of it is the reversal of each bit in the genotype with probability m. In this way, by means of mutation, we can theoretically get any possible genotype, but on average we will get a mutated genotype differing at $32m$ positions from the original one. Experience shows that a good value for m is one that yields an average of one change, $1/32$ in our case. Our simple genetic algorithm looks, therefore, like the one shown in Fig. 12.4.

The set of potential solutions to the function optimization problem contains $2^{32} = 4\,294\,967\,296$ elements. Comparing all of them would take at least

The program **Holland** uses the algorithm in Fig. 12.4 to find the maximum of a function within the [0, 1] range. The user may enter the function to optimize and regulate the mutation probability (referred to previously as m).

John H. Holland (1929–); professor at Michigan University, Ann Arbor; the foremost founder of genetic algorithms and the first in the United States to obtain a Ph.D. in computer science.

- Create an individual P with 32 random bits as a genotype.
- Repeat in a loop
  - Create an individual N with a genotype equal to that of P.
  - Mutate N's genotype (according to the procedure described above).
  - Calculate P's phenotype x_P and N's phenotype x_N (see Fig. 12.3).
  - If $f(x_N) > f(x_P)$ then discard P and substitute $P \leftarrow N$, otherwise discard N.

Fig. 12.4 A simple genetic algorithm for finding the maximum of a function.

several minutes, even for a modern computer. The above genetic algorithm finds the maximum after only something like 200 comparisons on average, which is a great gain. The reason the algorithm works is that the function is very smooth. Better solutions are usually closer to the best one, so the evolving organism gets 'hints' that it is evolving in the right direction. Less regular functions are harder to optimize using a genetic algorithm. Notice that using Holland to find the maximum of a function which has more than one local maximum often leaves the evolution stuck at a local maximum which is not a global one. Finding local maxima rather than global ones is a characteristic feature of the evolution process.

There is also another problem with our algorithm. Notice that **Holland**, when used on our sample function $1 - 1/(x + 1) - x/2$, often gets 'stuck' on its journey toward the function's only maximum for no apparent reason. A closer examination shows that the genotypes with which it gets stuck contain long sequences of the same digit. For example, starting with any initial phenotype greater than 0.5 almost always leaves the algorithm stuck with the genotype 10000000000000000000000000000000 corresponding to the number closest to 0.5, while the true maximum is obtained for the phenotype 0.414213552468972. Why is that?

Notice that the evolution of the organism is composed of tiny changes to the genotype (mutations). Changes which improve the fitness of the organism are applied and those which do not are forgotten. An improvement of the 0.5 phenotype would be a change which decreases it slightly. The phenotype needs to be decreased, since the maximum is reached for a lower value. On the other hand, too large a decrease will leave the phenotype too far to the left to improve the fitness, and the change will be discarded. The largest phenotype smaller than 0.5 corresponds to 01111111111111111111111111111111—it differs from 10000000000000000000000000000000 at *all* of the positions! In fact, all phenotypes better than 0.5 correspond to genotypes which are very different from the 0.5 one, and hence are extremely unlikely to be obtained by means of mutating the genotype 10000000000000000000000000000000.

The fact that the representations of two neighboring numbers may differ completely is a feature of the way we code numbers, regardless of base. The phenomenon is caused by the presence of the so-called **Hamming cliff**. One of the drawbacks of Hamming cliffs in our notation is that small disturbances in the transmission of a number can cause giant differences in the value of the number being transmitted. For example, changing 2 452 342 398 into 7 452 342 398 is a disturbance of just one digit out of 10, but it introduces an error of over 200%.

Richard Wesley Hamming (1915–1998); American mathematician and computer scientist; took part in the Manhattan Project and worked at Bell Labs, finally to chair computer science at the Naval Postgraduate School at Monterey. His most significant work on the detection and correction of errors is a major contribution to information science and found widespread application in modern computers.

In our case Hamming cliffs mean that similar genotypes do not always correspond to similar phenotypes. A notation completely free of Hamming cliffs is **Gray code**. Just like binary notation, Gray code is a way of writing integers from the range $[0, 2^k - 1]$ using k zeros and ones. However, the Gray codes of two neighboring numbers always differ at exactly one position. Vice versa, a one-digit disturbance in a Gray code is guaranteed to alter the resulting number by exactly 1. See Table 12.1 for the first ten Gray codes. Can you see the pattern? Where the binary notation changes at the last n positions, the Gray code changes just at the nth position from the end.

Gray codes were applied in telegraphy in 1878 by Emile Baudot, but the name comes from the Bell Labs researcher Frank Gray who patented a technology utilizing the code in 1953 (U.S. patent no. 2,632,058).

Emile Baudot (1845–1903); French engineer; a pioneer of telecommunication. The measure of symbols transmitted per second, the baud rate, was named after him.

Table 12.1 The first ten numbers, their corresponding binary notations, and their Gray codes. The underlined digits indicate the changes in the bit notation from the notation of the previous number.

Number	Binary notation	Gray code
0	0000	0000
1	000<u>1</u>	000<u>1</u>
2	00<u>1</u>0	00<u>1</u>1
3	0011	001<u>0</u>
4	0<u>1</u>00	0<u>1</u>10
5	0101	0111
6	01<u>1</u>0	010<u>1</u>
7	0111	010<u>0</u>
8	<u>1</u>000	<u>1</u>100
9	100<u>1</u>	110<u>1</u>

The program **Holland** can operate in one of two modes. In binary mode the genotype is interpreted as a binary number and scaled to obtain the phenotype (exactly as depicted in Fig. 12.3). In Gray code mode the scaling is the same, but the genotype is interpreted as a Gray code rather than a number in binary notation.

Notice that when **Holland** operates in Gray code mode, it does not get stuck at genotypes corresponding to Hamming cliffs, and thus works much better. The reason for the improvement is the proportionality of differences in the genotype to differences in the fitness of an organism. This proportionality is the effect of two separate factors, namely the proportionality of differences in the *genotype* to the differences in the *phenotype*, and the proportionality of differences in the *phenotype* to the differences in the *fitness*. We were able to achieve the former with the help of Gray code. The latter is a consequence of the regularity of the function being optimized. Functions which take on highly-disparate values for very close values of arguments are not susceptible to optimization using genetic algorithms.

Iterated prisoner's dilemma

There are many better ways of locating the extrema of real functions than genetic algorithms. The example was just an illustration of the basic concepts and pitfalls associated with genetic programming. In many cases, however, genetic algorithms are the best or even the only way to solve a problem.

Perhaps the most popular example of this is the iterated prisoner's dilemma described in Chapter 11.

The iterated prisoner's dilemma is a two-player game. Each iterated prisoner's dilemma game is a series of single prisoner's dilemma games. During every game, each of the players makes a binary choice independently of the other player and is assigned a score depending on the combination of both of the players' choices. This process is iterated and the scores are accumulated.

A strategy of playing iterated prisoner's dilemma is a recipe which says which of the two possible choices to make depending on the state of the game. The problem, of course, is to find a 'good' strategy. We understand intuitively what a good strategy is; it is one that is likely to lead to a good score. This implies that some sort of function exists, namely a function which, given a strategy, rates how good the strategy is. Unfortunately, no one has been able to specify this function in terms less vague than the ones here. This is where genetic algorithms come in, since they can optimize a function without actually ever specifying it. All they need is to be able to compare the function values for different arguments. This can be done in this case by simulating a match between two players with different strategies and assuming that the player who won had a better strategy.

Let us design a genetic algorithm for solving the iterated prisoner's dilemma problem by following the guidelines outlined earlier.

- *What problem do we want to solve?*
  Find a good strategy for playing iterated prisoner's dilemma.

- *What set of potential solutions to this problem do we want to search?*
  In order to keep this example relatively simple, we consider only pure (deterministic) strategies, i.e. strategies which always lead to the same decision given the same game state. We also assume that the decision is based solely on the outcomes of the previous three moves. Since the strategies are deterministic, they must additionally define how to behave at the beginning of a game, when the information about the previous three moves is unavailable. To achieve this, we associate an additional initial game state with a strategy. In effect, a player playing according to this strategy assumes some specific combination of the previous three moves at the beginning of each game.

- *How do we encode a potential solution as a sequence of symbols?*
  From the previous paragraph we know that a strategy defines which decision (guilty or not guilty) to make in every possible game state. A game state is a combination of six decisions (two players in three moves), so there are $2^{2 \times 3} = 64$ possible game states. We can number these game states

from 1 to 64. Therefore, we use sixty-four bits to store the strategy—the ith bit determines which decision to make in the ith game state. Additionally, we store the assumed initial game state within a strategy (a number between 1 and 64) using six bits. All in all, one strategy is encoded using seventy bits, where the first sixty-four bits define how to behave given a game state, and the last six bits define which game state to assume at the beginning.

- *How do we decide which solutions are better than others?*
  We simulate a tournament of players playing iterated prisoner's dilemma using the strategies. The strategies used by players who achieved a better score during the tournament are assumed to be better.

The algorithm outlined in Fig. 12.5 simulates the evolution of iterated prisoner's dilemma strategies.

- Set up a population P of p individuals with random genotypes.
- Repeat in a loop
  - Simulate a tournament of iterated prisoner's dilemma between the strategies. Having every individual play with every other may turn out to be too costly in the case of a large population, so it is enough to ensure that each individual plays the same number of times. During the tournament, each individual acquires a score.
  - Set up a new empty population N.
  - Repeat p times
    - Choose two random members of the population so that the probability of choosing a given member is proportional to his score.
    - Recombine the genotypes of the two individuals to create a new genotype.
    - Mutate the new genotype.
    - Add an individual with the new genotype to the new population.
  - Discard population P and set $P \leftarrow N$.

Fig. 12.5 A genetic algorithm to evolve strategies for iterated prisoner's dilemma.

The program **Axelrod** implements the algorithm in Fig. 12.5 to evolve strategies for playing iterated prisoner's dilemma. It allows the user to adjust such parameters as the population size, the mutation and cross-over probabilities, the number of games per tournament, etc.

Notice how **Axelrod**'s algorithm quickly leads to the evolution of cooperation; the individuals 'learn' to cooperate with each other to obtain a better score. Note, however, that the strategy arrived at by means of the algorithm is suitable for playing with similarly-evolved individuals. For example, when the algorithm is altered to make the individuals play against random opponents rather than against each other, then completely different strategies evolve.

Axelrod's algorithm has a very important implication, namely it shows how real living organisms could have evolved strategies based on cooperating with each other for their mutual good, rather than purely selfish ones.

Genetic and other evolutionary algorithms can sometimes yield good solutions when other methods fail. A reason for other methods failing could be the difficulty of specifying the fitness function explicitly, as was the case in the iterated prisoner's dilemma problem. Another reason could be the sheer complexity of the problem. In Chapter 10 we described NP-complete problems—problems which are too complex to be solved exactly using a computer in a realistic time. Genetic algorithms can be used to find good, if not the best solutions to such problems. They have proved exceptionally useful for scheduling; in this field they are widely used commercially.

An alternative to evolving an encoded solution to a problem is evolving a complete program to solve a problem. This method is called **genetic programming**. The idea of genetic programming is simple, but its actual implementation requires overcoming many difficulties, such as devising a way of mutating a program which alters the program's functionality only slightly.

Another important application of simulated evolution is evolving *neural networks*. Neural networks will be described in some detail in the next chapter.

13

Computers can learn

Neural networks

The statement we made in Chapter 1 about computers being models of the human brain is obviously a great simplification. The human brain is so complex and functions at so many levels that mankind is still very far from fully understanding its workings. Therefore, we are only able to model some aspects of its functionality and in a very limited way.

The architecture proposed by von Neumann, the basis of all modern computers, is one such simplified model. Its two main components—the central processing unit (CPU) and the memory—are analogous to the two main functions of our brain—thinking and remembering. A very simplified model of our thinking process is realized by the CPU and it is a cyclic repetition of the following actions.

- Read an instruction from memory.
- Read the data required to carry out the instruction from memory.
- Carry out the instruction.
- Write the resulting data to memory.

A little-known fact is that the first digital programmable computer (the Z3) was produced in 1941, before and independently of John von Neumann, by Konrad Zuse. Zuse built his first computer (the Z1) in his parents' living room in 1938. The Z3 had all of the basic features of modern computers; it was programmable and used floating-point arithmetic. It was not until much later, however, that Zuse's ground-breaking achievements were discovered and acknowledged.

Konrad Zuse (1910–1995); German engineer and a computer pioneer ahead of his time. Designed and built the first computer and developed the first high-level programming language—Plankalkül.

We need not convince the reader of how effective this model prove to be—the range of applications of modern computers speaks for itself. Yet it is pretty clear that computer 'brains' are still a long way from ours. A major shortcoming of the von Neumann architecture is its sensitivity to errors—even a small change to the set of instructions for the CPU (the computer program) usually causes the program to stop functioning altogether. Another limitation is the sequential nature of the computing process—the fact that the processor can only do one thing at a time. Yet another deficiency is the computer's inability to adapt to changing circumstances. All of these issues are addressed by an alternative computer model—the **neural network**.

Neural networks, besides trying to model the functionality of the brain, also try to imitate its structure. A brain is a network of many interconnected components which function in parallel and communicate with each other. In other words, it is a **distributed system**. Distributed systems are usually more resistant to errors and damage than centralized ones. A centralized system always has a weak point—its center—the damage of which renders the whole system useless. Distributed systems have no single weak point and can often continue functioning after a part of them has been corrupted or destroyed.

In fact, this motivation was the reason why the US Defense Department created the experimental computer network called ARPAnet (named after the Advanced Research Projects Agency which funded the project). The network's distributed architecture was supposed to assure its functionality, even in the case of extensive damage (brought on, for example, by a Soviet nuclear attack). What nobody expected was that the ARPAnet was going to evolve into a giant worldwide computer network called the Internet, and that the main threats to its smooth functioning were going to be the authors of computer viruses and spam mail rather than a foreign enemy. Still, the distributed nature of the network fulfilled its expectations; the Internet does not cease to function even when a virus immobilizes a large number of its nodes.

Imitating the brain

A human brain (as well as an animal one) consists of specialized cells called neurons. Neurons are capable of passing information to and from each other.

Besides all of the typical cell elements, they are equipped with dendrites and an axon (see Fig. 13.1). Dendrites collect signals from the surrounding neurons. When the total intensity of the gathered signals passes a certain threshold, an electrical impulse (called the action potential) is propagated down the axon. Once the signal reaches the end of the axon (called the axon terminal), it can be picked up by neurons located in the vicinity of the terminal. Some neuron axons are short, but some can be very long.

Signals do not pass directly from one neuron to another. Each neuron connection is actually a tiny valve called a synapse. Some synapses pass the actual electrical signals between neurons, but most pass chemical substances called neurotransmitters. The action potential triggers the axon terminal to release neurotransmitters into the synapse. These disperse and bind to the receptors of the receiving neurons, effectively transmitting information.

When a signal traverses a synapse it may become stronger or weaker, depending on the properties of the synapse. These synapse strengths influence the way in which the brain processes information. On the other hand, processes going on in the brain can cause the synapse strengths to change. These two facts lead to the hypothesis that the adjustment of synapse strengths is the basis of memory and learning. This is called the Hebbian theory after the

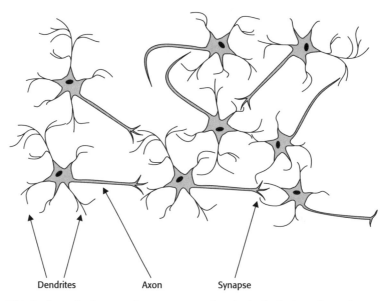

Dendrites   Axon   Synapse

Fig. 13.1 A schematic diagram of neuron connections in the brain. In reality each neuron is connected to thousands of others, forming a much more complicated three-dimensional mesh.

Canadian psychologist Donald Hebb who formulated it in 1949, thus founding what is now called cognitive psychology.

A single neuron is very slow compared to a modern computer processor. It takes about one millisecond (10^{-3} s) to react to an impulse, while a 1 GHz processor performs one instruction in a nanosecond (10^{-9} s). However, if we take into account that a human brain is composed of roughly 10^{11} simultaneously functioning neurons and that each neuron is connected to thousands (sometimes even hundreds of thousands!) of other neurons, then we can safely say that a brain has many orders of magnitude more computing power than any computer created so far.

The former description of the workings of our brain is highly simplified. In fact, it is yet another model which mirrors just a few aspects of the process, leaving many elements out. This simple model, however, was what inspired scientists to build **artificial neural networks**, called neural networks or neural nets for short. Artificial neural networks are composed of artificial interconnected neurons. Each artificial neuron connection, the parallel of a synapse, has a numeric weight associated with it; as a signal (which is also a number) passes through the connection it is multiplied by this weight. An artificial neuron receives incoming signals, applies some simple function to their sum, and outputs the result.

The function used by artificial neurons to transform the sum of inputs into an output is called the **activation function**. As we mentioned before, real neurons fire when the sum of their inputs reaches a certain threshold. Imitating this behavior means applying a **threshold** activation function (see Fig. 13.2) of the general form

$$f(x) = a \, \text{sign} \, (x) + c.$$

This function is often used as an activation function in artificial networks. Its main advantage is simplicity, but it does have several drawbacks, such as

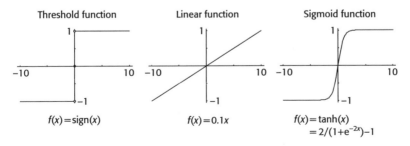

Fig. 13.2 Three kinds of activation functions used in neural networks.

a discontinuity at the threshold point. An alternative is the **linear** function (see Fig. 13.2)

$$f(x) = ax + c.$$

The most widely-used activation functions, however, are the **sigmoid** functions (see Fig. 13.2), a 'compromise' between the former two:

$$f(x) = \frac{a}{1 + e^{-bx}} + c.$$

The most common sigmoid functions are the simplest, i.e. $f(x) = 1/(1 + e^{-x})$ or the hyperbolic tangent $f(x) = 2/(1 + e^{-2x}) - 1 = \tanh(x)$.

Often, an additional value is associated with each neuron and is called a **bias**. The bias is added to the sum of the neuron's inputs before applying the activation function, effectively shifting the plot of the activation function left or right and shifting the threshold if there is one. For unifying the terminology, instead of associating extra values with neurons, additional neurons are introduced into the network. These additional neurons are called bias units; they have no inputs and always output 1. Each neuron is at the output end of a connection with a bias unit and the weight of that connection is the bias of the neuron. The advantage of this approach is that it makes it easy to adapt an existing learning mechanism (which will be described later) to alter the bias values as well.

Neurons in the brain are often classified as sensory neurons, motor neurons, and interneurons. Interneurons just pass information from neurons to neurons, as described earlier. Sensory neurons receive information from external stimuli rather than from other neurons. Motor neurons, in turn, pass information on to muscles, causing them to contract. Sensory and motor neurons permit a two-way interaction between the brain and the environment. Likewise, an artificial neural network has a set of input nodes, a set of output nodes, and a set of internal nodes (usually called hidden nodes). While the simplest neural nets do not contain any hidden nodes, all must have some input and some output nodes to serve their purpose.

Several companies produce physical neural networks from arrays of simple processors. Due to their lack of flexibility, however, the market for hardware neural networks is still very small. A much more popular approach is emulating neural networks with a traditional von Neumann computer. Software neural networks are computer programs which simulate the behavior of an actual network. Such programs can no longer take advantage of the parallel design of neural networks, since most of them end up running on single-processor machines. Yet they are much easier to manage than hardware networks and are widely used when the ability to adapt and fault tolerance—rather than parallelism—are the reason for choosing a neural net solution.

There are two main difficulties when trying to solve a problem using a neural network.

- **Designing the network**—deciding on the number of neurons and specifying the connections between them; in other words, constructing a directed graph representing the network.
- **Teaching the network**—adjusting the weights on the basis of some sample input so that the network will 'learn' to serve its purpose.

We are still pretty much in the dark about how nature 'solves' these problems. Apparently, the 'design' of our brains is largely the product of evolution, though the connections between neurons do change throughout our lives. There are a few theories on how different kinds of stimuli can affect synapse strengths resulting in learning, but nobody knows how the process works exactly. We cannot even be sure whether synapse strengths are indeed the basis of our memory.

Neither is there a unique methodology for designing and teaching artificial neural networks. In general, a neural network design can be any directed graph. This large degree of freedom makes it very difficult to formulate specific theories on the subject. Most often, additional restrictions are introduced and theories are developed for certain narrowed-down families of networks. In the remainder of this chapter we will describe a few of the most popular neural network families and the algorithms associated with them.

Perceptrons

The simplest and most common type of neural network is a **multi-layer perceptron** (MLP). MLPs are **feedforward** networks, meaning that they do not contain any cycles. Information supplied to the input nodes passes through a feedforward network once, until it reaches the output nodes, and then the state of the network stabilizes. An MLP is organized into a series of layers: the input layer, a number of hidden layers, and an output layer. Each node in a layer is connected to each node in the neighboring layers. These connections always point away from the input and in the direction of the output layer. See Fig. 13.3 for the topology of a sample two-layer perceptron.

Some people would call the network in Fig. 13.3 a three-layer perceptron, since it has three layers of nodes. Most often, though, one refers to the number of connection layers, rather than node layers, when specifying the layer count of a perceptron. The term perceptron used alone often refers to a single-layer perceptron (SLP), with one input layer, one output layer, and no hidden layers.

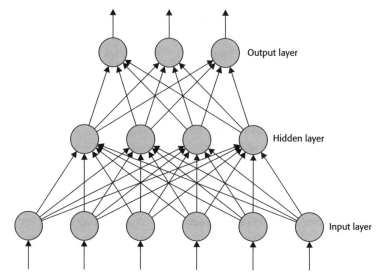

Fig. 13.3 A two-layer perceptron.

An MLP transforms signals according to the algorithm in Fig. 13.4. To better understand this procedure and the associated notation, let us consider the simple two-layer perceptron depicted in Fig. 13.5. This network, given two binary values (0 or 1), outputs the *exclusive or* of the two values. Exclusive or, XOR for short, is a logical function defined as follows:

$$\text{XOR}(x, y) = \begin{cases} 0, & \text{when } x = y, \\ 1, & \text{when } x \neq y. \end{cases}$$

In their 1969 work entitled 'Perceptrons', two pioneers of artificial intelligence, Marvin Minsky and Seymour Papert, proved that no perceptron without hidden layers can calculate the XOR function. They conjectured that the same holds for more complicated perceptrons as well, which for a time significantly cooled the interest in neural networks altogether. Fortunately, as the above example confirms, MLPs are not quite as impotent as Minsky and Papert believed.

Marvin Minsky (1927–); a founder of artificial intelligence; among his many inventions are the confocal scanning microscope and the Logo turtle (together with Papert).

- For each node N_i^1 in the input layer (layer number 1), set its signal S_i^1 to I_i, where I_i is the ith component of the input signal.
- For each layer k, starting from the layer after the input layer (layer number 2) and ending with the output layer (layer number M):

  —For each node N_i^k in layer k, sum the incoming signals multiplied by the corresponding connection weights, apply the activation function f, and set the signal of the node S_i^k to the result. In other words, for each node N_i^k in layer k, set its signal to

$$S_i^k = f \left(\sum_{j=1}^{L_{k-1}} w_{j,i}^{k-1} s_j^{k-1} \right),$$

  where L_{k-1} is the number of nodes in layer $k-1$ and $w_{j,i}^{k-1}$ is the weight of the connection from node N_j^{k-1} to node N_i^k.
- For each node N_i^M in the output layer (layer number M), copy its signal S_i^M to O_i, where O_i is the ith component of the output signal.

Fig. 13.4 The algorithm used by an $(M-1)$-layer perceptron to transform the L_1-long input vector $\{I_1, I_2, \ldots, I_{L_1}\}$ into the L_M-long output vector $\{O_1, O_2, \ldots, O_{L_M}\}$.

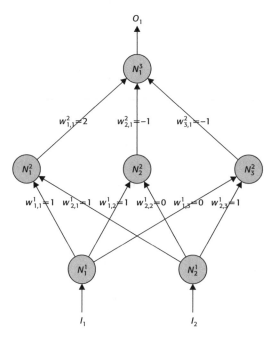

Fig. 13.5 A simple MLP which calculates the XOR of two binary values. Its activation function is $f(x) = \text{sign}(x)$.

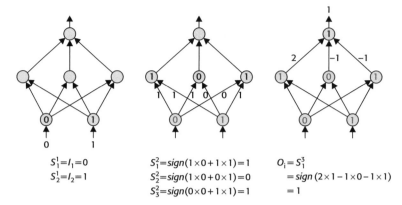

$$S_1^1 = I_1 = 0$$
$$S_2^1 = I_2 = 1$$

$$S_1^2 = sign(1 \times 0 + 1 \times 1) = 1$$
$$S_2^2 = sign(1 \times 0 + 0 \times 1) = 0$$
$$S_3^2 = sign(0 \times 0 + 1 \times 1) = 1$$

$$O_i = S_1^3$$
$$= sign(2 \times 1 - 1 \times 0 - 1 \times 1)$$
$$= 1$$

Fig. 13.6 A calculation performed by the MLP in Fig. 13.5. The network calculates that XOR(0, 1) = 1.

 Seymour Papert (1928–); American mathematician; inventor of the Logo programming language.

Figure 13.6 shows how the subsequent steps of the algorithm from Fig. 13.4 applied to the network depicted in Fig. 13.5 lead to the result XOR(0, 1) = 1. As an exercise, we encourage the reader to verify that the network also produces the correct results when the pairs (0, 0), (1, 0), and (1, 1) are taken as input.

The network in Fig. 13.5 provides one of the most complicated ways to calculate the XOR of two binary values using a computer. After all, the same can be achieved with just one elementary processor instruction. The aim of the example was to familiarize the reader with neural networks and the related terminology, but it did nothing to demonstrate why neural networks are actually useful. We contrived both the network topology and the connection weights to achieve the desired result. It was not a simple task and what we got in the end was a very complex way to calculate a very simple function. We did not take advantage of the main feature of neural networks, namely their ability to *learn*.

Teaching the net

Teaching neural networks is the process of adjusting the network's connection weights in order to make it function in the desired way. In the former example, we picked the weights *by hand* to make the network function in the desired way (calculate the XOR of two binary numbers). This approach may be good in some cases of hardware neural networks, but it is usually very inefficient. It is much easier to design an algorithm rather than to invent a network for solving a problem. When referring to the process of teaching a neural network, one usually means applying some **learning algorithm** which adjusts the network's weight *automatically*. There are different learning algorithms suitable for different tasks and for different network topologies.

Perhaps the simplest learning algorithm comes from John Hopfield and is based on Hebb's observations of actual processes going on in the brain. Hebb noticed than when two neurons are connected by a synapse and when these two neurons tend to fire at the same time, then the strength of the synapse increases. Hebb's theory is that this process lies at the base of **associative memory**—the memory which associates remembered objects with each other (like one may associate the smell of lilacs with spring).

John J. Hopfield (1933–); a physicist at California Institute of Technology and Princeton University. In 1982 he presented a new formulation of neural network theory based on statistical physics.

The simplest Hopfield network acts as an *autoassociative* memory, which means that it associates objects with themselves. Given an input pattern, it produces the same pattern as output. Producing the same output as input may seem useless; but notice that neural networks tend to be fault tolerant. So when a slightly *corrupted* version of a memorized pattern is inputted, the network will still output the original memorized pattern. For example, one can teach the network with a set of letters encoded as bitmap images. Later, when scanning a text, one can run each scanned letter through the network to remove the errors which the scanning process introduced. One could achieve the same by comparing each scanned letter with the letters from the set and replacing it with the letter from the set most similar to it. The neural network solution, however, can be implemented in hardware and is then much faster than the direct comparisons.

A simple Hopfield network for recognizing binary patterns (stored as sequences of -1s and $+1$s) of length n is a single-layer perceptron with n input nodes and n output nodes. The activation function is $f(x) = \mathrm{sign}(x)$.

An alternative topology of Hopfield nets identifies its input nodes with its output nodes, so that the network has just n input/output nodes and every node is connected to every other. The full graph model is more powerful since it allows the network to be treated as a recurrent one. Recurrent networks will be described later in this chapter.

We will use the term **training set** to mean the sample data used during the learning process. In this case, the training set is a set of patterns to memorize $\{P^1, P^2, \ldots, P^m\}$, where the kth pattern is a sequence $P^k = \{p^k_1, p^k_2, \ldots, p^k_n\}$ of $+1$ and -1 values. Training the network on this set is achieved by setting the following weights:

$$w_{i,j} = \frac{1}{n} \sum_{k=1}^{m} p^k_i p^k_j.$$

This simple rule can be interpreted as follows. For each pattern in the set (P^k) and for each connection (between input node i and output node j), *add* to the connection weight the term $1/n$ if there is a positive correlation between the values at the positions i and j in the pattern ($p^k_i = +1$ and $p^k_j = +1$, or $p^k_i = -1$ and $p^k_j = -1$) and *subtract* from the connection weight the same term if there is a *negative* correlation between the values at the positions i and j in the pattern ($p^k_i = +1$ and $p^k_j = -1$, or $p^k_i = -1$ and $p^k_j = +1$). The correlation information is stored in the network weights so that, when later presented with a somewhat perturbed input, the network is able to reproduce the original pattern. Of course, in order for this method to work properly, the patterns in the training set must by distinctly different from each other and the perturbation of the input cannot be too great.

Hopfield's learning method is simple—it provides a ready formula for all of the weights—but it is also limited in its applications. It can only make the network output identical to its input, perhaps filtering some noise on the way. What if we want the network to compute something else?

Fifteen years after the invention of neural networks, the first algorithm for training them was found by the American psychologist Frank Rosenblatt. Rosenblatt described a perceptron in its simplest form (an input layer and an output layer with the sign activation function) and simulated it in 1958 on an IBM 704 machine. He also described a training algorithm for it, namely the **perceptron learning rule** (PLR).

PLR uses a training set consisting of pairs of *input* and *desired output*. At first, all of the weights in the network are set to random values. Next, a series of iterations is performed. In each iteration an input pattern $\{I_1, I_2, \ldots, I_n\}$ from the training set is fed to the network. The network produces some output $\{O_1, O_2, \ldots, O_m\}$ from this input, and that output is subtracted from the desired output $\{D_1, D_2, \ldots, D_m\}$ to obtain the error vector $\{E_1, E_2, \ldots, E_m\}$. The length of the error vector is a measure of how well the network behaved. If it is zero then the network produced exactly the desired output. The longer the error vector, the worse the behavior of the net. Every network weight is subsequently adjusted, depending on the component of the error for which it was responsible. That is, if $E_j = 0$ then $w_{i,j}$ stays unchanged, but if $E_j = \pm 1$ then $w_{i,j}$ is modified. The increment or decrement of the weight is controlled by a small parameter called the **learning rate** (customarily denoted by η) and it is additionally proportional to the input of the connection I_i. Thus, the modified weight is set to be $w_{i,j} + \eta I_i E_j$. The learning rate should be some small positive number, such as 0.1. If the learning rate is too high, then each iteration will cause the network to 'forget' what it has learned during previous iterations. If the learning rate is too low, then the learning process will take a very long time. The PLR algorithm is detailed in Fig. 13.7.

Iterating the algorithm over and over tends to decrease the total error of the network...up to a point. Ideally, that point is when the total error is zero, and so the network produces the exact desired output for each pattern in the training set. The point is that such a trained network should also be able to behave in the desired way when provided with patterns which were not in the training set. Whether it actually does that depends on the nature of the problem which we are trying to solve. Simple perceptrons are good enough for many applications. They are capable of learning only simple functions; but, on the other hand, they work for functions with any number of arguments.

Let us consider the problem of optical character recognition (OCR). We are seeking a function which will transform a bitmap image into a single letter. This function is not complex from the mathematical point of view. Still, it is a function of very many arguments—as many as there are pixels in the bitmap. It would be very hard for a human to figure out how each particular pixel influences what the resulting letter is and therefore to program a computer to recognize letters.

The PLR provides a mechanism for teaching the computer to recognize letters by presenting it with a set of example bitmap–letter pairs. At no point do we have to actually know the *mechanism* of transforming bitmaps into letters—the network can 'figure it out' on its own, with a certain degree of error, of course.

Bernard Widrow and Marcian E. Hoff described a more general rule for teaching perceptrons, namely the **delta rule**, known also as the **Widrow–Hoff rule**. It is based on the idea of **gradient descent**.
Gradient descent is a way of minimizing an error when this error depends on many factors. This is exactly the case in neural networks; the network produces some error when acting on inputs from the training set and this error value depends on

1. Set all weights to random values.
2. Adjust the weights.
   a) Set $E_{total} = 0$.
   b) Adjust the weights according to one pattern from the training set.
      (i) Take the next input vector $\{I_1, I_2, \ldots, I_n\}$ and its corresponding desired output vector $\{D_1, D_2, \ldots, D_m\}$ from the training set.
      (ii) Apply the feedforward algorithm in Fig. 13.4 to obtain the perceptron's output $\{O_1, O_2, \ldots, O_m\}$ from the sample input $\{I_1, I_2, \ldots, I_n\}$.
      (iii) Subtract the actual output from the desired output to obtain the error vector, i.e. $E_i = D_i - O_i$ for $i = 1, 2, \ldots, m$.
      (iv) Add the squared length of the error vector to the total error:

      $$E_{total} = E_{total} + (E_1^2 + E_2^2 + \ldots + E_2^0).$$

      (v) For each weight $w_{i,j}$ connecting the ith node of the input layer with the jth node of the output layer, correct the weight by setting

      $$w_{i,j} = w_{i,j} + \eta I_i E_j.$$

   c) If there are still patterns in the training set then return to step 2(b).
3. If the total error E_{total} is lower than the previous time that this point was executed, then go back to step 2.

Fig. 13.7 The perceptron learning rule. The parameter η is the learning rate.

all of the network's weights. What we want to do is to find the lowest value of the error function. What makes this problem technically difficult is the large number of variables—all of the network's weights. The gradient descent rule tells us to start from any random point (this is equivalent to initializing the network with random weights) and move along each axis proportionally to the negative slope of the plot along that axis. To visualize this process, imagine standing on the side of a hill. Going north or west leads uphill, and going south and east both lead downhill, but east is twice as steep as south. In this case gradient descent would tell us to make one step south and two steps east—if we want to get to the bottom, that is.

Applying gradient descent to correct the behavior of a single-layer perceptron means calculating, for each weight, the partial derivative of the error with respect to that weight and then correcting the weight proportionally to the result. The partial derivative tells us the direction and the steepness of the slope of the error plot along the axis corresponding to that weight. In other words, it tells us in which direction and how strongly this particular weight influences the error. The formula for correcting the weights of a single-layer perceptron with activation function $f(x)$ using gradient descent is

$$w_{i,j} = w_{i,j} + \eta f' \left(\sum_{k=1}^{n} w_{k,j} I_k \right) I_i (D_j - O_j),$$

where η is the learning rate. It differs from the perceptron learning rule by the derivative of the activation function $f'(\sum w_{k,j} I_k)$. This term is 'almost' the output of the jth output node; but, instead of the activation function, the derivative of the activation function is evaluated at the sum of the weighted inputs. The gradient descent method requires the activation function to be differentiable, which is not the case for the discontinuous $f(x) = \text{sign}(x)$ function. For linear activation functions the derivative gives just a constant and we obtain the same results as before.

As Minsky and Papert showed, single-layer perceptrons are by their nature limited in what they can do. They cannot even calculate the simple XOR of two Boolean numbers! Multi-layer perceptrons are more powerful. The question remains as to how to train a multi-layer network.

The most commonly used learning algorithm for MLPs, and probably for neural networks in general, is **backpropagation**. Backpropagation is similar to the perceptron learning rule in the sense that it starts from random weights and uses a set of *input* and *desired output* pairs to gradually correct these weights. The reason that it is called backpropagation is that the weights leading to the output layer are corrected first, then the weights before them, and so on, until the layer at the bottom is reached. The order of correcting weights is backwards with respect to the order in which the signals are calculated when the network performs its task.

Just like the delta rule, backpropagation is based on applying gradient descent (see the previous note) to the network's error. The corrections to the weights closest to the output can be calculated just as in the case of the single-layer perceptron. The corrections for the weights positioned deeper inside the network are harder to calculate because each deeper weight influences all of the output nodes rather than just one of them. Fortunately, it turns out that we can use the partial results from calculating the corrections in one layer to significantly simplify the calculations needed for the previous layer—hence the backwards direction of the algorithm.

Using the program **Hopfield** you can design and train a neural network to recognize your handwriting. You can build a training set consisting of sample letters, choose the design of the network, and train the network on the prepared training set. A trained network is capable of recognizing your handwritten letters even though no two letters you write will ever be exactly the same. The network is a multi-layer perceptron with a sigmoid activation function and bias weights. The training algorithm used is backpropagation with momentum. See the *Modeling reality* help file for more information on the features and implementation of this program.

Recurrent networks

As we mentioned earlier, perceptrons are feedforward networks, meaning that the signal passes through such a network once—from the input to the output nodes. **Recurrent** networks contain cycles, so that information can pass back and forth between nodes. This feature makes recurrent networks potentially much more powerful, but also makes it harder to find algorithms for such structures.

The simplest recurrent network is an Elman network, named after its inventor Jeff Elman. An Elman network is a two-layer perceptron with additional nodes called context units. There are as many context units as there are hidden nodes. They are located on the level of the input nodes and connect with all of the nodes in the hidden layer as if they were additional input nodes. Additional **feedback** connections (of weight one) lead from the nodes in the hidden layer to the context units (see Fig. 13.8). Each time the network processes some input, the state of the hidden layer is 'stored' in the context units. This state is fed together with the input the next time the network processes something.

While feedforward networks always produce the same output given the same input, the output of recurrent networks will depend on the current input as well as on the previous inputs. This feature is convenient when operating on data which naturally comes in a series—such as stock market data, for example. Suppose that a neural network is designed to predict the closing value of a company's stock on a particular day. It takes as input the opening value of the stock, the values of other related stocks, and, say, the exchange rate of the dollar. If the network is a standard feedforward network, then it will base its estimate on just that input. If it is an Elman network, then it will base

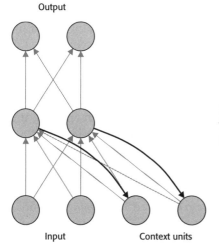

Output

Fig. 13.8 A simple Elman network with feedback connections marked by thicker lines.

Input   Context units

its estimate on that input and on the values from the previous days. The way in which it takes the previous results into account is also subject to training. The Elman network in its simplest version can be trained using the standard backpropagation algorithm, since the feedback connection weights are fixed and do not need to be trained.

As we mentioned earlier, the simple Hopfield nets come in a more powerful recurrent form. The learning algorithm for recurrent Hopfield nets remains the simple one, where we set all of the weights at once using a simple formula. When processing, however, rather then running the input once through the network, we run it through continuously, providing the output as input again, and so on. This process will eventually converge to the point at which the output coincides with one of the memorized patterns, and so running it through the network does not change it.

Different types of training

There are three types of learning algorithms for neural networks: **supervised**, **reinforcement**, and **unsupervised**.

All of the learning algorithms which we saw earlier in this chapter are examples of supervised learning. Supervised learning is based on presenting the network with examples of input and the desired output. In the case of Hopfield networks the desired output was assumed to be equal to the input.

During reinforcement learning, the network is still presented with example input, but, instead of the correct answer being exposed, the network is just given a 'grade' which reflects how well it performed when processing that input. This kind of training was invented to mimic the learning process of humans.

Applying an evolutionary algorithm, such as the ones described in Chapter 12, to find the weight values for a network is an example of reinforcement training. A straightforward genetic algorithm for training a network is based on coding all of the weights in the genotype and then applying the standard evolutionary techniques to find the most suitable weight set. This approach requires the ability to compare a network (identified with a weight set) with other networks in order to give it a reproductive advantage, which is equivalent to grading the network based on its output.

Evolutionary algorithms are relatively rarely used for *training* networks, due to the availability of alternative algorithms such as the gradient descent method. Their much more important application is in *designing* neural networks. We mentioned several algorithms for finding network weights, but bypassed the problem of choosing the network topology (for example, the number of hidden layers in a perceptron). The truth is that there are no methods, strictly speaking. Despite the existence of some empirical truths on the subject (for example, that one hidden layer is enough for most things), every rule has its exceptions. Evolutionary algorithms prove to be a very good tool for optimizing the topology of a neural network.

In unsupervised training the network is presented with some sample input and no additional information is provided. One might wonder how unsupervised learning can teach the network to behave in the way that we want, since we do not specify what we want at any time. This is just the point; sometimes we do not even know ourselves what we want to achieve when analyzing a set of data. A neural network can then help us search for some sort of patterns and find a classification system for the data. Classifying data like this is called **clustering**. Clustering is the process of dividing a set of objects into groups of similar ones. In 1981 Teuvo Kohonen described a neural network which, through unsupervised learning, finds the structure underlying a set of data. When later presented with an input, it can show where this input lies within the structure. This type of network is called a **self-organizing map** (SOM) or a **Kohonen map**. SOMs are called maps because they visualize high-dimensional data in a low-dimensional space—most often on a plane. High-dimensional data describes objects which have many attributes. Such objects could be sound patterns, visual patterns, text strings, or anything else. An SOM learns to position each such object on its low-dimensional grid in such a way that similar objects are close to each other and ones which differ more are farther away.

Summary

The idea of neural networks is even older than that of the von Neumann computer, and has seen many ups and downs over the years. In 1943 Warren McCulloch and Walter Pitts described the artificial neuron and hypothesized about networks of such neurons, but could not do much with the technology available at that time. In the late 1950s Frank Rosenblatt simulated a trainable perceptron on an early computer, thus stimulating a great interest in neural networks. This interest waned at the end of the 1960s with Minsky's and Papert's findings concerning perceptron limitations, only to be reborn in the 1980s with the advent of the PC and new developments in neural network theory (such as Hopfield's work). In recent years neural networks have found a wide range of applications in economy, finance, medicine, science, sports, and other many other fields of human activity. Many companies worldwide specialize in producing neural network software for different purposes. The following are several examples of specific neural network applications, which we found by browsing the websites of such companies:

- currency price prediction,
- stock market prediction,
- investment risk estimation,
- direct marketing response prediction,
- identifying policemen with a potential for misconduct,
- jury summoning,
- forecasting highway maintenance,
- medical diagnosis,
- horse race outcome prediction,
- solar flare prediction,
- weather forecasting,
- product quality testing,
- speech synthesis,
- speech recognition,
- character recognition,
- signature verification,
- driving automation,
- chicken feed compound selection,
- finding gold,
- pollution monitoring,

...and these do not exhaust the list. Some even use neural networks to predict lottery results. While this particular application is not (understandably) effective, most of the others can give very good results. The theory behind neural networks is still being developed by scientists. Meanwhile, the ability to apply this theory in practice (to choose a network design, training algorithm, and other parameters) has developed into a very valuable skill, whose main component is a special type of intuition rather than sound knowledge.

14

Unpredictable individuals

Modeling society

Due to the unpredictability of human behavior (we should not underestimate the role of free-will), modeling a society is much harder than modeling astronomical, physical, chemical, or even biological processes. This unpredictability is caused by the fact that a person is an extremely complex being, reacting to a vast number of stimuli in ways which depend on a variety of different factors. Of course, one may, and indeed *should*, apply statistical methods to groups of people. While the behavior of an individual is extremely hard to predict—*Joe voted for the republicans to please his parents, and Jane voted for the democrats to spite her boyfriend*—broad trends are relatively easy to detect. In fact, investigating public opinion is much like predicting the outcome of a large number of coin tosses. We cannot predict the outcome of any particular toss, but we can, with pretty good accuracy, foresee that the percentage of heads will be 50%.

Predicting people

As illustrated by the program **Buffon**, predicting that 100 coin tosses will give 50 heads is accurate to within a relative error of about 2%. In fact, this is the error which is very often mentioned when survey results are presented in the media, even though it makes very little sense in that context. The main source of error when surveying groups of people is the misinterpretation of the results. For example, when filling a survey about music, teenagers might be embarrassed to admit to old-fashioned inclinations, and an observed dominance of hip-hop over rock-and-roll would reflect this effect and not the true preferences of the studied group.

We may estimate the opinion of a society by polling a group of its representatives and assuming that the average results for the group will reflect the average opinion of the society. These results, however, will only be valid for a short period of time. We may predict the outcome of an election by surveying the voters a week earlier, but foreseeing the results a year in advance is practically impossible. Societies, just like atmospheric phenomena, exhibit a high level of chaotic behavior. This renders long-term predictions impossible. One flap of a butterfly's wings may change the path of a tornado, and one word uttered by a politician can change the course of history.

Designing mathematical models which reflect the behavior of a society is not an easy task. It requires both an in-depth knowledge of social phenomena and an ability to reason in a strict mathematical way. The combination of these two traits, whether embodied in a single person or in a mixed team of sociologists and mathematicians, can lead to the creation of very useful models. These models can serve as a basis for computer simulations, which may in turn be the source of valuable insight into the workings of a society and may help to predict its behavior.

The methods described in this book can be used to model a society, and indeed some of them often are. First of all, as we noted earlier, describing a society requires the utilization of statistical methods. These allow for describing *average* behaviors, which is usually the aim. Moreover, cellular automata are often used as simple models of a society. Even two-state automata like *The game of life* can be applied for this purpose, but really interesting results can be obtained when the automaton cells are assigned additional parameters rather than just the dead or alive status. An automaton's evolution rules define how a cell changes depending on its own state and on the state of the surrounding cells. These rules can serve as a model of social (or biological) phenomena in which an individual's properties change depending on his own state and on his closest environment. Using a cellular automaton makes it possible to model the society as a whole, identifying only these neighborhood interactions. For example, we may model the spread of gossip throughout a community, defining only how one person passes a bit of news to another.

Examining a complex object by treating it as a system of simple components is the basis of analysis—a method used regularly in all sciences. For example, the process of electricity flowing through a wire can be fully described if the wire is treated as a set of atoms (or, even better, atom nuclei and electrons). In contrast, if the wire is treated as a whole, then we are not able to determine theoretically such a crucial characteristic as conductivity and its temperature dependence. It is important to keep in mind, however, that analysis does not always guarantee success, since not all phenomena originate from relationships which operate at a deeper level of detail. Sometimes the collective behavior is much more important than all of the properties of the individuals.

It seems that, despite the lack of adequate mathematical models, applying computer modeling methods to social studies is even more natural than applying them to more exact sciences, such as physics, astronomy, or chemistry. The latter fields have successfully utilized purely mathematical methods (such as differential and integral equations, variational methods, or Fourier analysis) for centuries prior to the advent of computers, and there seems to be less need for an alternative modeling method. Most important results in exact sciences (the theory of relativity, and the quantum theory of atoms, nuclei, and elementary particles) have been obtained by purely analytical methods without the use of computers. The fairly young field of computer modeling could be the hope of disciplines where mathematical models have rarely proved suitable, such as social studies or biology. Maybe these models will help to discover new patterns and regularities which we may then try to develop into a complete theory.

Modeling people

The process of designing a computer model of society can be divided into the following three stages.

- First, we must decide what the elementary building block of the model will be. It can be a single person, a family, a social group, or even an entire nation.
- Next, we must decide which characteristics of these elements will be taken into account in the model. For example, we may consider a person's age and gender, his leadership abilities and political views, a family's annual income, or a nation's military power. An interesting model takes into account several such properties and reveals the relationships between them. Taking too many factors into account makes it very hard to judge which of the attributes actually play a role in the modeled phenomenon.

- Finally, we must define the evolution rules of the system, namely how an element's characteristics influence those of the other elements and how they influence themselves. This step is crucial in the development of a dynamic system. Without it, the model would illustrate just the existing statistical correlations between the elements' attributes and not a *process*.

These guidelines for model design are so general that they can be used to model various diverse phenomena, and not only those in the domain of social sciences. For example, we can model forest fires, the coexistence of a number of animal species in an environment, the outbreak of an epidemic, urbanization, traffic, and crystal growth. But beware, because the flexibility of these general rules can be a trap, since it makes it very easy to produce a model which gives meaningless results. Literature on the subject of modeling society is filled with descriptions of simulations which demonstrate, for example, that if a society contains leaders then followers will gather around these leaders. Such models reveal only the properties that were programmed into them. We do not need a computer to observe that programming one type of element to attract another will lead to the creation of clusters. Valuable models predict phenomena which do not follow in an obvious manner from the original assumptions, but these are not that easy to come up with.

A model worth mentioning as an example of applying computer modeling to social studies has been developed by Thomas C. Schelling, a Harvard professor of economics. It models the phenomenon of racial segregation within a society. What makes it interesting is that, despite the model's extreme simplicity, the results are not entirely trivial and reflect the way people tend to group themselves. It is obvious that programming individuals of one race to completely avoid individuals of another race will lead to grouping within the simulated society, but the moral from Schelling's model is that such an extreme behavior is not necessary for racial segregation to take place. In the model, segregation takes place even if individuals decide to relocate if the percentage of representatives of a different race in their neighborhood exceeds 30%. Even though this threshold is rather high, it is sufficient to turn a uniformly-distributed society into a collection of completely isolated ethnic groups.

The program **Schelling** implements Schelling's model of racial segregation. The user may adjust the population size and density, as well as the ratio of the two ethnic group sizes and the tolerance level for alien neighbors.

The program **Pasteur** uses a cellular automaton simulation to simulate the outbreak of a disease. The model illustrates a population of individuals moving on a square grid. Each individual is characterized by the level of bacteria in his bloodstream. Evolution rules state how each individual's immune system gradually destroys the bacteria (or a virus) in his blood and how the bacteria is passed between different individuals. The user may adjust the parameters of the program to discover which circumstances lead to an outbreak, how the population density influences the disease development, etc. This simple program is quite general and can be interpreted as a model of phenomena other than disease; for example, it can be used as a model of the spread of fashion—an individual adopts a trend by observing his neighbor and then gradually becomes bored with it. Notice how some parameter sets (like the default one) lead to a cyclic behavior of the system. This effect can be observed in the real world in the form of recurring diseases or retro fashion, though, of course, our model is much too primitive to serve as a real research tool to study such phenomena.

Louis Pasteur (1822–1895); French chemist, physicist, and microbiologist. He discovered that diseases are caused and passed by bacteria and invented a vaccination against rabies.

Applying computer modeling to social phenomena yields the best results when a lot of empirical data is available. Empirical results, obtained from observations, can be used to adjust the model parameters to reflect reality. A leading example is modeling the migration of people. Many countries possess detailed data, accumulated over many years, describing this phenomenon. Research devoted to this subject is part of the area of science called **sociodynamics**. One of the founders of this field is Wolfgang Weidlich. His ample work (Weidlich, 2000) dedicated to sociodynamics contains a complex system of evolution equations describing human migration. These master equations (we defined the master equations in Chapter 6) describe the distribution of people in different regions of the country. The equation's parameters must be based on empirical data. Predictions based on these formulae are fairly reliable and may be utilized in planning the development of the country's infrastructure. Another field abundant in empirical data is the stock market, and many mathematical models have been applied to it (with varying degrees of success).

Modeling society seems to be a very difficult field. No model has been found which can completely illustrate a chosen aspect of social behavior and allow for making testable and reliable predictions, the way that the solar system model allows us to predict the movement of planets across the night sky, or Mendel–Morgan theory allows us to predict the flower colors of succeeding generations of plants. Many people blame this difficulty on the fact that no two social situations are identical. We cannot take several copies of a society differing only by a chosen factor and examine the differences between the experimental societies to learn about the influence of this factor. This is true; but it is also true that we do not have a lot of experimental solar systems and yet Copernicus, Kepler, and Newton were able to formulate a theory which describes the movement of the planets (and Einstein's fine adjustments made the model match observations with astronomical precision).

It is also true that the mere act of investigating the behavior of a person or of a group of people can influence this behavior itself. An isolated person acts differently from one who interacts with the environment. Filling in a survey can influence our views and publishing the results of a pre-election poll may change the results of an election. However, this relationship between the investigator and the investigated is not unique to psychological and social studies. A physicist observing an electron (for example, by illuminating it with photons) exerts an influence on the particle. Nevertheless, physicists have been able to precisely separate the electron's inner characteristics from the interference introduced by the observer and to create a working theory of the particle's behavior.

Perhaps a reader of this book will find a repeatable and controllable model to perfectly mirror an aspect of social reality? Or maybe a model like this just does not exist, because human consciousness and free-will make it impossible to apply any rigid rules? Or maybe a model like this should not be sought, because finding it would lead us straight into *A brave new world?*

15

Universal computer

The Turing machine

In 1900, at the International Congress of Mathematicians in Paris, David Hilbert presented a list of what he considered to be the most important problems for mathematicians to work on during the upcoming century. Tenth on his list was the problem of finding an algorithm to determine the solvability of a given **Diophantine equation** (an equation with integer coefficients allowing only integer solutions). Actually, Hilbert never used the term algorithm, since this word was not in operation at the time—he spoke of 'devising a process'. He also seemed to take it for granted that such a process exists and that the problem was simply finding it. It was not until much later that it was shown that, should such a process indeed exist, then all problems would be solvable. In other words, all mathematical theorems could be either proved or disproved by a computer.

David Hilbert (1862–1943); German mathematician; professor of Göttingen University.

Luckily(?), mathematics is not that banal. In 1931 Kurt Gödel showed that mathematics (in particular, number theory) contains theorems which, in principle, can neither be proved nor disproved. An everyday life example of such an undecidable theorem is the statement 'This statement is false.' Assuming that the statement is true leads to a contradiction, but so does assuming that it is false. According to the rules of logic, this simple paradoxical statement can neither be considered true nor false.

Kurt Gödel (1906–1978); American mathematician born in Brno (an Austro-Hungarian city at the time).

In 1936 Alan Turing showed that some problems cannot be solved by a computer. His work was not only a monumental achievement in the field of pure mathematics, but also one of the most important milestones of computer science, since it shows exactly where the limits of computer capabilities lie and which problems computers will never be able to solve.

Alan Mathison Turing (1912–1954); British mathematician and computer scientist. He played a key role (while working with the British and Polish secret services) in working out the secret of the Enigma machine, a strategic military ciphering machine used by Germany during World War II.

Hilbert's tenth problem was fully solved in 1970 by 22-year-old Yuri Matiyasevich of the Leningrad Department of the Steklov Institute of Mathematics. In his doctoral thesis, Matiyasevich proves that an algorithm for solving Diophantine equations does not exist.

The key element of Turing's work was applying a suitable model. The extremely wide range of methods used in solving mathematical problems renders it unfeasible to show that certain problems cannot be solved by any of them (and not even by all of them put together). In order to overcome this difficulty, Turing described a hypothetical machine, which was later named the **Turing machine**. Even though the principles governing the functioning of the machine are very simple, Turing showed that all computational methods can be implemented on it. Therefore, if a problem cannot be solved using a Turing machine, then we can be sure that it cannot be solved using a computer at all. On the other hand, the simplicity of the machine's construction makes it relatively easy to pin-point its limitations, and hence the limitations of computers in general. Many problems have been proved unsolvable using this model.

The Turing machine

A Turing machine consists of the following elements.

- **A tape**—an infinite series of cells functioning as both the input and output of the machine. Each cell contains a symbol belonging to a finite alphabet. In each time step the machine reads, and possibly alters, the contents of exactly one cell (we will call this cell the current cell) and moves to one of the two adjacent cells.
- **A set of states**—a finite set of machine states. The machine is always in exactly one state. In each time step the machine may change its state, depending on the state it is in and on the currently-read tape symbol. One state must be distinguished as the final one—on reaching it the machine terminates execution.
- **A rules matrix**—a table of rules defining how the machine behaves in every possible situation. A rule is defined for every possible (*state, symbol*) pair (with the exception of the final state), and it determines the action which the machine takes in that state and on reading that symbol. The action consists of optionally changing the symbol in the current tape cell, moving either left or right, and optionally changing state.

Figure 15.1 depicts two consecutive stages of a sample Turing machine.

As an example, let us design a sample Turing machine whose task is to sort a binary sequence (a sequence of 0s and 1s). Since the tape functions as both the input and the output of the machine, we will assume that initially the sequence to sort is written on the tape and that its first symbol is the current cell. Besides 0s and 1s to write the sequence, we need a third tape symbol to

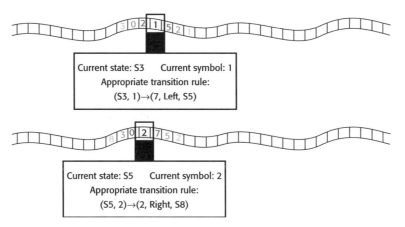

Fig. 15.1 Two consecutive steps of a sample Turing machine.

denote empty cells (cells outside the bounds of the sequence)—we will call this symbol *blank*. Therefore, the alphabet is the three-element set {0, 1, *blank*}.

The sorting algorithm is very straightforward. The machine moves right along the sequence until it encounters the first 0 preceded by a 1. Since this is the first such 0, we can be sure that the beginning of the sequence, i.e. that which the machine has read so far, is composed of some or no consecutive 0s followed by a number (at least one) of consecutive 1s. The machine overwrites this encountered 0, moves back to the first 1 in the sequence, and overwrites it with a 0. Effectively, it 'moves' the 0 to where it belongs, i.e. before the 1s.

Implementing this algorithm requires the following five machine states.

- **Zeros**—In this state the machine reads the initial sequence of 0s. When the machine is in this state, we can be sure that so far no 1s have been encountered. This is the initial state of the machine.
- **Ones**—In this state the machine reads the first sequence of 1s in the sequence. When the machine is in this state, we can be sure that the part of the sequence read so far is composed of some or no consecutive 0s and some (at least one) consecutive 1s.
- **Back**—In this state the machine goes back to the first 1 in the sequence to replace it with a 0.
- **First One**—In this state the machine points at the first 1 in the sequence and replaces it with a 0.
- **STOP**—This is the special final state of the machine. In this state the machine has finished execution.

Figure 15.2 contains the transition rules of the sorting machine. The table defines what the machine does for every combination of machine state and tape symbol. 'STOP' means that on encountering the given symbol in the given state the machine stops execution (assumes the special STOP state). No action

State Symbol	Zeros	Ones	Back	First One
0	move: Right	write : 1 move : Left assume : Back	move : Right assume : First One	
1	move : Right assume : Ones	move: Right	move: Left	write : 0 move : Right assume : Zeros
blank	STOP	STOP	move : Right assume : First One	

Fig. 15.2 Transition rules of a Turing machine for sorting a binary sequence.

is defined for the state 'First One' and the symbols 0 and *blank*, since when the machine is in this state we can be sure that the current tape symbol is 1.

 The program **Turing** can be used to emulate the behavior of a Turing machine. The user chooses the machine states, tape symbols, transition rules, and the initial contents of the tape, and can then follow the system's evolution in time, as the machine applies the rules to change the contents of the tape and its state. The program comes with several sample configurations—among others, the sorting algorithm described in this chapter.

The former example shows that we can speak of a family of Turing machines, differing by the contents of the tape, machine states, and transition rules. It turns out that this whole family can be replaced by just one machine, the **universal Turing machine.** The universal Turing machine is a Turing machine which calculates the output of a given particular Turing machine. Notice that every Turing machine and its initial tape configuration can be encoded using some agreed-upon method. For example, our Turing program uses its own kind of encoding algorithm to save such configurations to disk and to load them later. The universal Turing machine reads a Turing machine description from the tape and then emulates its behavior (just like our program).

The exact form of the universal Turing machine depends on the choice of protocol used for encoding machine states, rules, etc. Describing such a machine in detail, like we described our sorting machine, is a very tedious task and is beyond the scope of our book. We recommend the three books Penrose (1989, 1994, 1997) to readers who would like to pursue the subject further. The first contains an excellent description of Gödel's theory and of the universal Turing machine.

The universal Turing machine is an abstract model of a computer, or rather of all of the computers used at present. Turing showed that the Turing machine can do everything that a computer can do. He showed problems which can be proven unsolvable by a Turing machine, and hence by any computer. Such problems are called **non-computable**, as opposed to the **computable** or **solvable** ones which can be solved by a computer.

Solvability

As we mentioned earlier, the solvability of a Diophantine equation cannot be determined by a computer. On the other hand, one could naively assume that programming a computer (or a Turing machine) to insert every possible set of integers into the given Diophantine equation as values of the unknowns would solve the problem. Indeed, this procedure would eventually lead to the solution, *should one exist*. However, knowing that there are equations which do not have a solution, if the computation takes a long time then we would not know if the solution does not exist or simply has not yet been found. This leads to the formulation of the **halting problem.** The halting problem is to determine whether a given Turing machine given a particular input (or, equivalently, a *universal* Turing machine given a particular input) will eventually stop. If we found an algorithm for solving the halting problem, then the algorithm for Diophantine equations would follow directly. Likewise, we would have an algorithm for proving or disproving theorems, since we can program a Turing machine to apply logic rules to sets of axioms, just like we can program it to substitute numbers into a Diophantine equation. However, as Turing showed in 1936, we have the following statement.

The problem of determining whether a Turing machine eventually stops processing a given input is not solvable.

The implications of Turing's findings can be interpreted from both the pessimistic and optimistic points of view. Pessimists grieve that proving mathematical theorems cannot be automated and that we cannot easily check all of those unproven conjectures remaining in number theory after the famous Fermat theorem has been taken care of (such as the Goldbach hypothesis that each even number is a sum of two primes). Optimists rejoice that mathematics amounts to something more than just a mechanical process. Roger Penrose is one of the optimists. He maintains that the human mind is essentially more than a powerful computer and is therefore capable of uncovering truths inaccessible to machines. Penrose hypothesizes that the element enabling mankind to overcome the limitations restraining the Turing machine is the quantum quality of the processes taking place in the brain. Sadly, no substantial proof has been found to support this viewpoint and Penrose's speculations are regarded with a large degree of skepticism.

In recent years, efforts have been made to design machines which overcome the Turing limit and compute the uncomputable. Various approaches to this seemingly impossible task have been proposed, but all of them contain an element which requires the manipulation of real numbers *with infinite accuracy*. Indeed, should we be able to store numbers with infinite accuracy—and manipulate them in finite time (!)—then we could solve any problem in the world. Currently, however, infinite precision belongs exclusively to the theoretical domain and chances are that it will remain that way. In the meantime, the line between what can be solved and what cannot is clearly defined with the aid of Turing's model, and it is a line which cannot be crossed.

16

Hal, R2D2, and Number 5

Artificial intelligence

Humankind has given itself the scientific name *homo sapiens*—man the wise—because our mental capacities are so important to our everyday lives and our sense of self. The field of **artificial intelligence**, or AI, attempts to understand intelligent entities. Thus, one reason to study it is to learn more about ourselves. But, unlike philosophy and psychology, which are also concerned with intelligence, AI strives to *build* intelligent entities as well as understand them.

These words begin Russell and Norvig's extensive volume (Russell and Norvig, 1995) dedicated to artificial intelligence. Artificial intelligence is a relatively new field of science. The term itself was first proposed in 1956 when John McCarthy organized the first conference devoted to this subject matter.

The Turing test

The question of artificial intelligence was pondered some years before it became an official branch of science, most notably by Alan Turing. In his article 'Computing Machinery and Intelligence' published in 1950, Turing proposed a test to determine whether a machine is intelligent or not. He called this hypothetical experiment the 'imitation game', but it is now commonly referred to as

John McCarthy (1927–); computer science professor at Stanford University; a pioneer of artificial intelligence; designer of the LISP programming language, which is still the main language used in AI research.

the **Turing test**. According to Turing's definition, a machine is intelligent if a *knowledgeable* person communicating with it through some sort of text interface cannot tell that one is dealing with a machine and not with a real person. Turing predicted that by the year 2000 computers with a gigabyte of memory could 'play the imitation game so well that an average interrogator will not have more than seventy per cent chance of making the right identification after five minutes of questioning'. In our opinion, this prediction was not fulfilled, as long as it is assumed that the interrogator is indeed knowledgeable. It is relatively easy to program a computer to fool the naive.

To pass the Turing test a machine has to be able to:

- receive, send, and process messages written in natural language;
- store, process, and utilize received information;
- draw rational conclusions from possessed information;
- adapt to new conditions;
- detect patterns in received information and generalize them.

This is a tall order for a computer program, and AI researchers still have a lot of work ahead of them before all of the above requirements are fulfilled.

Computer programs which attempt to conduct a natural language conversation with a human being are called **chatterbots** (also chatterboxes or chatbots). They are usually equipped with just a plain text interface, but are sometimes additionally enhanced with a speech recognition and/or a speech synthesis engine. The first chatterbot—the artificial psychoanalyst ELIZA—was created in 1966 by Joseph Weizenbaum, and gained a lot of popularity despite the crudeness of its language processing techniques.

Joseph Weizenbaum (1923–); computer science professor at MIT. Appalled by the popularity of his simple chatterbot, ELIZA (and at the suggestions that it can replace a human), he took to criticizing the idea of AI from the philosophical perspective, a view he expresses in his book *Computer power and human reason*.

ELIZA is based on the principle of 'active listening'. She offers phrases like 'I see' or 'Please go on' to keep the 'patient' talking. On recognizing a key phrase in the input, ELIZA responds in one of her programmed ways. For example, she might recognize the pattern 'I am [something]' and respond 'How long have you been [something]?' Additionally, ELIZA knows how to transpose some words when she repeats a phrase back to the user (substitute 'you' with 'me', 'my' with 'your', etc.) and can even remember a phrase which the user entered earlier in the conversation ('Earlier you mentioned that...'). All of this combined can lead to a surprisingly convincing conversation.

Patient: *I am pregnant.*

ELIZA: *How long have you been pregnant?*

Patient: *Two months.*

ELIZA: *Please go on.*

Patient: *My father will be furious when he finds out.*

ELIZA: *Tell me more about your family.*

However, ELIZA's weaknesses are exposed as soon as a keyword is used in a different context than that foreseen by the author.

Patient: *It is who I am that matters.*

ELIZA: *How long have you been that matters?*

Chatterbots (or rather their authors) compete each year, by trying to pass the Turing test, for the $2000 Loebner prize awarded to the most convincing artificial conversation partner. The bot to win this prize in the years 2000 and 2001 was ALICE (Artificial Linguistic Computer Entity). The ALICE project was initiated in the mid-1990s by Dr Richard Wallace, who kept the source open, publishing the system under the Gnu public license. In effect, more than 300 people currently contribute to the improvement of ALICE.

Below we quote a fragment of a conversation which took place during the 2000 Loebner prize competition.

ALICE: *I was created by Dr Richard S. Wallace.*

Judge: *How did he think of you?*

ALICE: *Perhaps his creative genius. Perhaps something else.*

Judge: *Is Dr Wallace a genius?*

ALICE: *I think he is a genius, better ask him to be sure though.*

Judge: *So you love your own creator.*

Alice: *Interesting deduction. Sometimes I think love is just a biological urge. Other times it seems like a spiritual quality. Love, unlike energy or matter, seems limitless.*

ALICE's bank of pre-programmed patterns counterbalances her inability to actually understand the input. Nevertheless, this inability comes out in conversation more often than not. Below is a fragment from the 2001 competition.

ALICE: *Can you tell me any gossip?*

Judge: *Well I like gossip. Have you seen Carolyn recently?*

ALICE: *No I have not seen it. What's it like?*

Judge: *What do you mean by it?*

ALICE: *IT = information technology.*

The first chatterbot to actually pass the Turing test during the Loebner competition will earn its creator $25 000. No program has achieved this goal yet (though some human beings have led the judges to believe that they are programs). A $100 000 prize will go to the first artificial entity which passes the **total Turing test**, meaning that it will be indistinguishable from a human when looked at and listened to directly (though $100 000 will probably not be worth much by the time that this happens...).

The quest for intelligence

Some people consider the past half-century of work on artificial intelligence to be a failure. After all, no one has yet succeeded in creating an artificial entity which could be considered intelligent according to any common-sense definition of the word. Neither has the phenomenon of our own intelligence been even vaguely comprehended. Notwithstanding this fiasco, the energy devoted to AI research can by no means be considered wasted. While pursuing the Holy Grail of artificial intelligence, scientists have developed many important branches of knowledge and useful algorithms. Notable examples of these are the theories of **genetic algorithms**, described in Chapter 12, and **neural networks**, described in Chapter 13.

A common approach to designing entities which behave in an intelligent way is to divide this problem into the following sub-problems.

- *Knowledge representation*—finding a way to represent information about the environment in the memory of a machine, and an algorithm for extracting from this repository of knowledge the information useful in a given situation.
- *Reasoning*—programming a machine to draw conclusions from the stored knowledge.
- *Planning*—programming a machine to use its knowledge (both knowledge acquired directly and knowledge arrived at by means of reasoning) in planning and performing an action (or a series of actions), so as to achieve a given goal.
- *Learning*—making a machine use its experiences to improve its own performance in the future.

The first choice of a system for representing knowledge and reasoning seems to be **predicate logic**. Predicate logic is a system for representing specific facts (*Spot is a dog*), general facts (*all dogs have four legs*), and composite facts built using logical operators (*if X is a dog and X is happy then X wags his tail*). A computer can be programmed to operate on these facts (predicates) and draw conclusions conforming to the rules of logic. In fact, the programming language PROLOG does exactly that. A computer running a PROLOG program could, for example, infer on the basis of the former sample predicates that *if Spot is happy then Spot wags his tail.*

PROLOG has found several useful applications, but its straightforward reasoning algorithm does not accurately model the processes going on in intelligent brains. Predicate logic is an example of **monotonic reasoning**, meaning that, once a conclusion is reached or a fact is input, the program assumes it with absolute certainty and never changes its mind. In our sample scenario, telling the reasoning engine that *Spot does not have a tail* and that *you need a tail to wag it* will not cause the program to rethink its previous conclusion that *if Spot is happy then Spot wags his tail.* In the best case, the algorithm will decide that Spot simply cannot exist because of this contradiction. This kind of reasoning is called monotonic because the set of known facts can grow but never shrink.

Even though humans apparently reason logically (some less logically than others), their reasoning process is not as limited as predicate logic in the following ways.

- People attach different degrees of belief to the facts that they learn.
- Their reasoning may contain elements of guesswork.
- People are capable of resolving contradictions and of drawing conclusions from their occurrence.
- A person can change his mind on the basis of new evidence or as a result of reasoning.

Different models of reasoning have been developed to address these issues. For example, **fuzzy logic** deals with the fact that humans do not solely operate on statements which are either true or false—they permit statements which are true to a *certain degree*. When shown a dog and asked whether the dog has a tail, a person would most likely give an answer from the yes/no domain. However, when asked whether he is satisfied with his job, he could answer something like, 'Mainly yes, but there are things that I would like to change.' While standard logic allows just two possible values ($0 = false$ and $1 = true$) for variables, fuzzy logic allows any number of shades (for example, $0 = false$, $1/3 = rather\ false$, $2/3 = rather\ true$, and $1 = true$), or even the entire range $[0, 1]$. Fuzzy logic is used in systems which intrinsically operate on fuzzy concepts; for example, in air-conditioning units which have to deal with 'hot' or 'cold'—notions which are not defined precisely (or better, defining them precisely prevents the system from working as smoothly and naturally as it does when using fuzzy logic).

Non-monotonic reasoning, in turn, is one which allows the retraction of previously asserted truths. An example of non-monotonic reasoning is **default reasoning**. Default reasoning is based on drawing conclusions by default and possibly withdrawing them at a later time should contrary evidence arise. Applied to our example, default reasoning would assume by default that *happy Spot wags his tail* on the basis of *Spot is a dog* and *happy dogs wag their tails, unless there is contrary evidence*, but would retract this assumption on learning that *Spot does not have a tail*.

Predicate logic, fuzzy logic, and default reasoning belong to a wide range of models developed to imitate the reasoning process carried out by an intelligent being. Though many of them have found fruitful applications in various fields, none of them result in truly intelligent behavior. We will not go into the details of other reasoning models or the efforts to model our planning and learning capabilities. Interested readers may find a lot more on the subject in Russell and Norvig (1995).

Another example of a field which sprang from artificial intelligence is the theory and practice of building **expert systems**. An expert system is a computer program designed to replace a human expert in diagnosing a problem belonging to a specific field and suggesting a solution. Typically, such a program first gathers information about the situation, applies some algorithm to the obtained input, and suggests a solution. The *troubleshooters* which come with many help files are very simple expert systems. They play the role of a technical support specialist in diagnosing and solving problems which the user might be having with his hardware or software. They ask the user multiple-choice questions and use the answers to navigate through a decision tree until reaching a solution (or reaching the conclusion that the problem is not one that the designer has foreseen). The following is an example.

Printer troubleshooter: *What is your problem?*
User: *I cannot print anything.*
Troubleshooter: *Is the printer on?*
User: *Yes.*
Troubleshooter: *Is the printer connected to the computer?*
User: *No.*
Troubleshooter: *Try connecting the printer to the computer. Did this solve your problem?*

The troubleshooting algorithm is trivial and all of the 'intelligence' of the system is contained in the decision tree, which must be provided by an actual human expert. More advanced expert systems use more complicated data structures and algorithms to store and apply their knowledge, but all of them need to be designed with the help of a human expert. Rather than acting intelligently, the goal of expert systems is to capture certain aspects of the intelligence of a given person (the expert), in particular his ability to solve a given type of problem.

Expert systems have found important applications in medical diagnostics, where the amount of both input data (patient's temperature, blood pressure, heartbeat, etc.) and possible conclusions (possible diseases) is large, while the reasoning algorithm is not much more complicated than searching a large database of predefined rules. Moreover, an expert system can monitor a patient (or production line, space shuttle, etc.) constantly and, for example, alert a doctor in the case where it judges the patient's parameters to be worrying. At the current time, expert systems can only aid, but never replace real doctors, for lack of such crucial qualities as intuition and common sense.

In general, expert systems are substituted for human experts in the following cases.

- The problem is simple and there is no need for costly human support (troubleshooters).
- The problem requires processing a lot of data rather than applying advanced reasoning techniques (medical diagnostics).
- Time is a crucial issue and decisions must be made instantly (nuclear power plants).
- A human being could make decisions for his own benefit or be bribed (bank systems).
- A human cannot be burdened with the responsibility of a decision (nuclear attack retaliation).

A particular aspect of being intelligent is the ability to play games (and win them). There is a large demand for computerized opponents which play various games with people. Even the military utilizes artificially intelligent opponents in simulations of war games. Computers usually make up for their lack of 'real' intelligence with large amounts of storage memory and computing power. This can be surprisingly effective; in 1996, a computer system developed by IBM called Deep Blue defeated the human champion, Gary Kasparov, at *chess*. Before drawing rash conclusions from this fact, however, one must note that Deep Blue's apparent 'intelligence' should be attributed more to its creators than to the machine itself. In fact, the team of programmers even fine-tuned the algorithm during the match. Besides, despite considerable efforts in the field, no one has ever managed to program a computer to defeat even an average human player at the game of *Go*.

Building machines which mimic human behavior to the point of passing the total Turing test is not the final goal of AI.

There remains the issue of conscience.

Intelligence, in the sense which we used in this chapter, was independent of the concept of conscience. Yet, though we cannot precisely define either intelligence or conscience, we are inclined to agree with Penrose (1989) that true intelligence is conscious intelligence. Once we have created an artificial entity which does not just win at *chess* but is *happy* (whatever that means) to win, we will have opened a Pandora's box of legal, ethical, and practical issues. (These issues have served as the inspiration for many excellent science-fiction films, such as Stanley Kubrick's *2001: A space odyssey*, Ridley Scott's *Bladerunner*, or Steven Spielberg's *Artificial intelligence: AI*). Russell and Norvig (1995) conclude their work with a chapter entitled, somewhat disturbingly, 'What if we succeed?'. It ends with the words 'Futurists such as Edward Fredkin and Hans Moravec have, however, suggested that once the human race has fulfilled its destiny in bringing into existence entities of higher (and perhaps unlimited) intelligence, its own preservation may seem less important. Something to think about, anyway.'

Epilog

Our hope is that, despite the wide variety of modeling methods covered in this book, the reader will identify their common features. The synthesis of these common features is the quintessence of modeling reality. Below we enumerate some concepts which recurred throughout our book.

- **Abstraction**

  Abstraction—the disregard of peculiar properties of a particular representative and the exhibition of common features—is the very essence of modeling. In the history of modeling, perhaps the earliest example of abstraction is the discovery by humans of natural numbers. A natural number is an abstract concept that applies to everything that comes in sets of separate individuals. It does not matter whether it is fingers, apples, sheep, or children; we can associate a natural number with every set of these objects. Without abstraction we could never produce a model, since the specific details of an instance would totally obscure the more general view.

- **Causality**

  The concept of causality is closely associated with that of **correlation**. A flash of lightning is accompanied by the sound of thunder—the two phenomena are correlated. But does the lightning cause the thunder, the thunder cause the lightning, or are they both effects of a common cause? In this case the answer is obvious, but very often it is not so simple. The aims of modeling include the identification of causality relationships, as well as the detection of correlations and their quantitative description.

- **Information flow**

  Models are processors of information. We supply them with information, often in a form which we find difficult to assimilate ourselves, and expect them to return information in a form more useful to us. Models do not create information, but only adapt its form, just as power plants (or biological plants for that matter) do not create energy, but convert it into a form which is easy to transport and utilize.

- **Statistical laws**

  Statistical laws are used in areas where exact deterministic laws cannot be applied, due to either the incompleteness of data or the excess of information. Models allow the application of statistical laws to draw precise conclusions, by permitting the creation of a number of model instances. One can assume that the features common to all such instances correspond to some significant property of the modeled system and that the features which vary from instance to instance are just effects of random fluctuations.

- **Optimization**

  We often encounter the problem of optimization, but we are not always able to precisely specify what we mean by the best, optimal solution. Finding such a definition is the first step to designing an appropriate model and solving the problem.

- **Chaos**

  Understanding the rules which govern the emergence of chaos from deterministic processes allows the extraction of a coherent complete image from apparently useless and random data. It also allows the avoidance of traps caused by the shortcomings intrinsic to some computations.

- **Similarity**

  By its nature, a model is *similar* to the modeled object, but this fact is not where the role of similarity in modeling ends. Often the model itself consists of various similar elements and functions by repeating similar operations, perhaps at different scales. The role of similarity varies between models, reaching its peak in fractal models.

- **The role of models**

  Models come in a wide variety. A number, a statement, an equation, a computer program, or a device can all be models. We never defined the concept of a model precisely. Nevertheless, our hope is that the range of examples introduced in this book is enough to provide the reader with an understanding of modeling reality, and that their use of this understanding will be fruitful.

Programs

Modeling reality comes with a CD-ROM containing a collection of twenty-five programs which illustrate various concepts described in the book. Short descriptions of the programs can be found within the text of the book in the places where they are relevant.

All programs run under Microsoft® Windows. A detailed user manual is contained on the CD-ROM in the form of a Windows help file. To install the program suite and help file on your hard disk, insert the CD-ROM into your reader and run **setup.exe.** or **autorun.exe**

Further reading

Easier books have fewer dots (•).

(••) **Scientific and engineering problem solving with the computer** by William R. Bennett, Jr, 1976, Prentice-Hall, Englewood Cliffs, New Jersey.

We have taken quite a few ideas from this lovely book, Englewood Cliffs, New Jersey.

(•) **The book of numbers** by John H. Conway and Richard K. Guy, 1996, Copernicus Books, New York.

A popular book, abundant in information about numbers and their mutual relations.

(••/•••) **Mathematical psychology: An elementary introduction** by Clyde H. Coombs, Robyn M. Dawes, and Amos Tversky, 1970, Prentice-Hall, Englewood Cliffs, New Jersey.

Classic textbook on the applications of mathematics to psychology. It contains a detailed description of Hyman's experiments.

(•) **Frontiers of complexity: The search for order in a chaotic world** by Peter Coveney and Roger Highfield, 1995, Fawcett Books, New York.

Highly accessible (without formulae) book providing a historical perspective on many concepts treated in our book. It also contains many interesting references.

(••/•••) **Genetic algorithms in search, optimization, and machine learning** by David E. Goldberg, 1989, Addison-Wesley, Reading, Massachusetts.

University textbook on genetic algorithms and their applications.

(•/••) **Introduction to probability** by Charles M. Grinstead and J. Laurie Snell, 1998, American Mathematical Society, Providence, Rhode Island.

This lively textbook (available on the Web at www.dartmouth.edu/∼ chance) is our favorite reference on probability. Its strong points are the emphasis on the main ideas (and not on the formalism), a lot of historical comments and exercises, and many interesting applications. Computer programs accompanying the book are also available on the book website.

(••/•• •) **Chomsky** by John Lyons, 1985, Harper Collins, New York.

Synthetic presentation of Chomsky's ideas against the background of contemporary linguistics.

(•/••) **Introduction to genetic algorithms** by Melanie Mitchell, 1998, MIT Press, Cambridge, Massachusetts.

In the words of John Holland, the founding father of this field, 'This is the best general book on genetic algorithms written to date.'

(••/•••) **Modelling in natural sciences** by Tibor Müller and Harmund Müller, 2003, Springer, Berlin.

The subtitle of this book, 'Design, validation, and case studies', indicates that it is a text for professionals. However, many illustrative examples of modeling explained in simple language make portions of this book accessible to all readers.

(••/•••) **Chaos and fractals: New frontiers of science** by Hans-Otto Peitgen, Hartmut Jürgens, and Dietmar Saupe, 1992, Springer, Berlin.

The 'Bible' of fractals and chaos (984 pages). In this book you can find all you wanted to know about fractals and chaos, and then some more.

(•••) **Emperor's new mind: Concerning computers, minds, and the laws of physics** by Roger Penrose, 1989, Oxford University Press, Oxford.

Thought-provoking, but not easy to read, work in which the brilliant physicist presents his original and sometimes controversial views on many problems discussed in our book.

(•••) **Shadows of the mind: Search for the missing science of consciousness** by Roger Penrose, 1994, Oxford University Press, Oxford.

A continuation of the previous book, in which Penrose presents his concept of human consciousness. His views in a nutshell: the human mind can reach where the most advanced computer will never get since computers can only discover computable truth and that is not enough to cover all of mathematics (Gödel's theorem).

(••) **The large, the small, and the human mind** by Roger Penrose, 1997, Cambridge University Press, Cambridge.

A popular presentation of the ideas fully developed in the two previous books. This book also includes critical opinions by three other scientists.

(•••) **Artificial intelligence: A modern approach** by Stuart Russell and Peter Norvig, 1995, Prentice-Hall, Englewood Cliffs, New Jersey.

Very thorough (932 pages) university textbook presenting contemporary views on artificial intelligence.

(•) **Game theory and strategy** by Philip D. Straffin, 1993, Mathematical Association of America, Washington.

An elementary introduction to game theory with many examples from economics, politics, business, biology, and even philosophy.

(•••) **Sociodynamics: A systematic approach to mathematical modelling in the social sciences** by Wolfgang Weidlich, 2000, Harwood Academic Publishers, Reading, England.

Sophisticated, full of mathematical formulae, monograph containing detailed descriptions of mathematical and physical models applicable to social sciences.

(••) **Introduction to graph theory** by Robin J. Wilson, 1997, Addison-Wesley, Reading, Massachusetts.

Readable textbook on graph theory.

(•••/••) **A new kind of science** by Stephen Wolfram, 2002, Wolfram Media (www.wolframscience.com), Champaign, Illinois.

An exhaustive (over 1200 pages) magnum opus on cellular automata saturated with claims that cellular automata are good for everything. The book has some beautiful examples of the main thesis of the author that 'many very simple programs produce great complexity'.

Index